AF881137

The Army Learning Concept, Army Learning Model:

A GUIDE TO UNDERSTANDING AND IMPLEMENTATION

By Brent A. Anders, Ph.D.

The Army Learning Concept, Army Learning Model:

A GUIDE TO UNDERSTANDING AND IMPLEMENTATION

By Brent A. Anders, Ph.D.

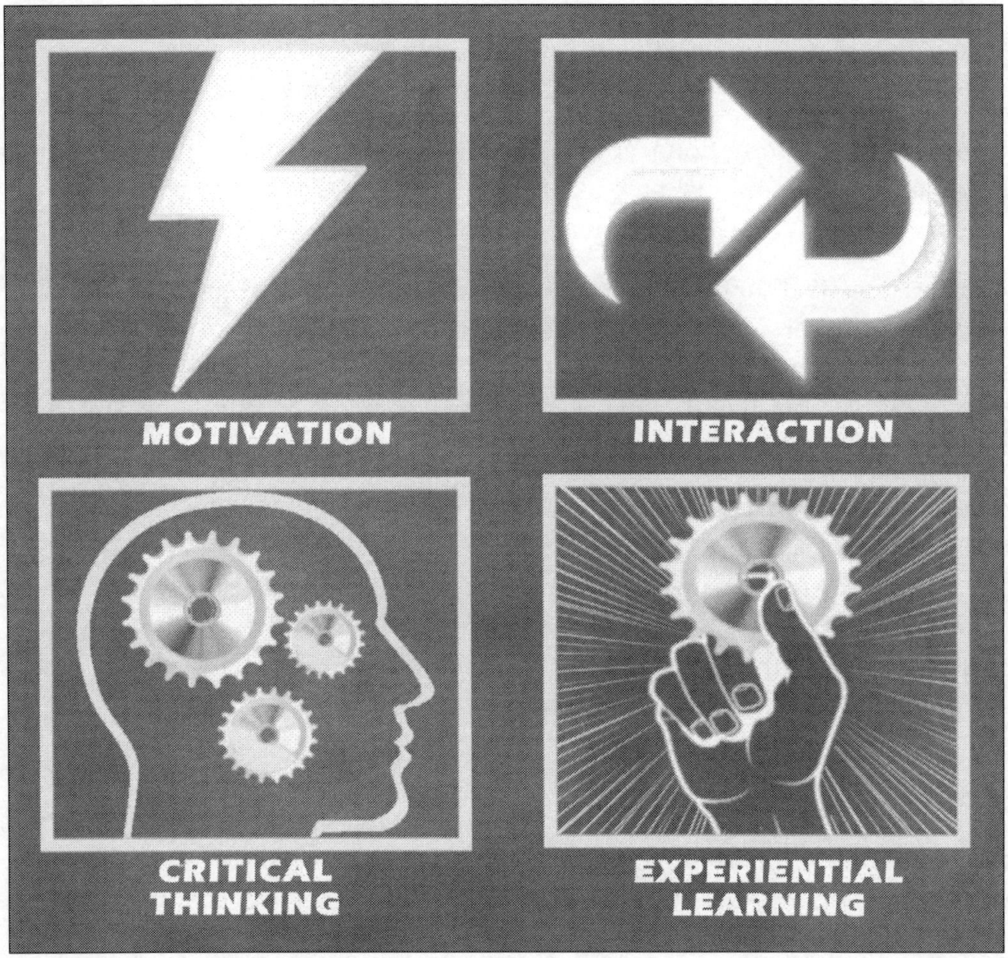

Sovorel Publishing

The Army Learning Concept,
Army Learning Model:
A GUIDE TO UNDERSTANDING AND IMPLEMENTATION

© Copyright 2019 by Brent A. Anders, Ph.D. All Rights Reserved (except where noted)

Published in the United States of America by Sovorel Publishing Company,
www.sovorelpublishing.com

This book is protected by international copyright law. No portion of this book may be reproduced without permission from the publisher, except as permitted by U.S. copyright law. For special permissions contact: Contact@SovorelPublishing.com

Note that websites described in this book may have disappeared or changed since this book was published. While the author and publisher have diligently prepared this book, neither of them make any warranties or representations as to the accuracy or completeness of the contents of this manuscript after publication due to issues beyond their control. The strategies and advice contained within this book may not be suitable for all situations. Neither the author or the publisher shall be liable for any loss of profit or any other type of commercial damages.

Although this book deals with the U.S. Army's Learning Concept, Army Learning Model, this is not an officially created Army manual or training circular. Brent A. Anders, is a currently serving Sergeant Major in the U.S. Army National Guard, but has written this book as a civilian and not as an official spokesperson for the U.S. military. The information presented is the author's personal perspectives/research from serving and teaching in the U.S. Army for over 25 years.

ISBN-13: 978-0-9987637-3-6 (digital), ISBN-13: 978-0-9987637-2-9 (paperback)

Keywords/Tags: Army Learning Concept, Army Learning Model, motivation, interaction, critical thinking, creative thinking, experiential learning, instructional methodologies, hands-on learning, instructional presence, online, distance, face-to-face, video, education, training

Cover, photos, and illustrations for this book created and/or edited by Brent A. Anders

The Army Learning Concept, Army Learning Model: A Guide to Understanding and Implementation © Copyright 2019 by Brent A. Anders, Ph.D. All Rights Reserved.

Contents

DEDICATION

Dedicated to all the brave men and women in all of our armed services. May your training always be relevant, helpful, motivating, impactful, and a positive educational experience. Thank you for everything that you do and for your many sacrifices. God bless you and your families.

ACKNOWLEDGEMENTS

First and foremost, I wish to acknowledge and publicly thank my beautiful wife Nelli Karapetyan Anders. Without her support and eagle eye editorial help this manuscript could not have been completed. I also wish to thank the U.S. Army and Kansas National Guard for providing me with multiple educational experiences and opportunities. And of course, I wish to thank God for his many blessings and giving me the ability and resources to complete this book.

FOREWARD

This book was specifically written for all educational practitioners, from junior Noncommissioned Officer (NCO) team leaders wanting to give a superior class to their squad, to certified Army instructors wanting to improve their official military course, to seasoned civilian professors wanting to enhance instruction at the university level. The information within this book provides a full comprehensive explanation of the Army Learning Concept (ALC), Army Learning Model (ALM) on both an academic and practical level. This book is designed to help those in the field of education to not only understand ALM but to actually use and implement it to improve instruction and learning. Information for this manuscript was put together from over 225 different academic and military sources and my own experiences spanning over 25 years in the fields of both higher academia and military instruction.

To assist in navigating and understanding ALC/ALM a detailed table of contents and index is provided. The Army Learning Concept and Army Learning Model are fully defined and explained in chapter one. Subsequent chapters expound upon the four main overarching concepts of the Army Learning Model: Motivation, Interaction, Critical Thinking, and Experiential Learning. Explanations, illustrations, and implementation examples are provided to fully understand and be able to use ALM at all levels (both within military and civilian education environments).

To gain the most out of this book we must accept the fact that although we might be a subject matter expert, a high ranking official, or a Ph.D. holder in our particular field, that does not in itself make us a great instructor, teacher, trainer, professor, or facilitator. We must also fully grasp and understand that there is a big difference between teaching (the process of instructing to help others understand and increase their capabilities) and learning (the transformation of one's knowledge/skills/attitudes – ways of thinking). Both deal with education but there are different systems involved in each (it is more than just data transfer and acquisition). The Army Learning Concept acknowledges these differences and addresses each via enhanced educational methodologies, focusing on the learning process, as well as being student-centered/learning-centered to maximize understanding, skills acquisition (both physical and mental capabilities), and long-term retention of knowledge gained.

CHAPTER 1: Army Learning Concept/Army Learning Model

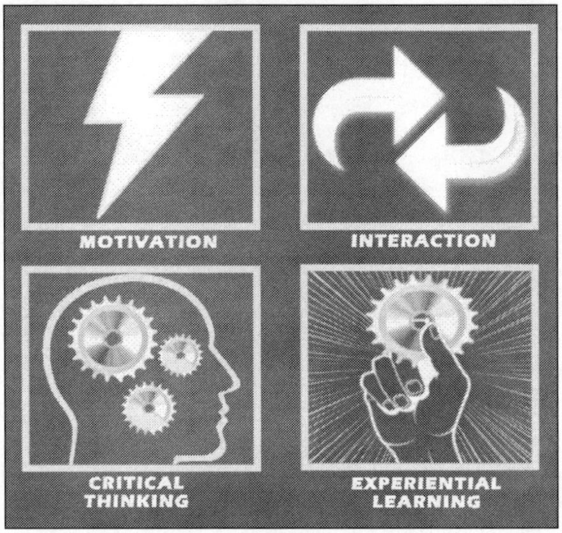

- The Army Learning Concept defined/explained
- The Army Learning Model defined/explained
- How to conduct instruction using ALM along with a logical brake down of ALM into its four main concepts: Motivation, Interaction, Critical Thinking, and Experiential Learning.

THE ARMY LEARNING CONCEPT

The Army Learning Concept (ALC) is the Army's ongoing learning environment that is adaptive, life-long, and provides a continuum of educational opportunities to provide knowledge and skills to soldiers/Army civilians so they can continually improve and succeed under ambiguous/chaotic conditions (Training and Doctrine Pamphlet 525-8-2, 2017). The Army Learning Concept was designed to house an Army-wide model for learning that incorporates the operational factors of full-spectrum operations, adaptability, decentralization, mastering of fundamentals, culture/language, and capitalization on students' experiences (Training and Doctrine Pamphlet 525-8-2, 2011, p. 11). Specific fundamentals of the Army Learning Concept are its adherence to adult learning principles (such as a need to know why-*relevancy*, a readiness to learn, and use of prior experience), student-centered instruction, and experiential (hands-on) learning whenever possible (Training and Doctrine Pamphlet 525-8-2, 2017).

The ALC is designed to improve educational effectiveness and move away from deprecated instructional methodologies that have persisted within the Army and periodically resurface due to time constraints and an educational force that isn't fully informed (Ferguson, 2014). A prime example of an outdated instructional methodology would be the basic lecture with an improperly created and administered slideshow without any type of interaction (Ferguson, 2014; Training and Doctrine Pamphlet 525-8-2, 2011). This type of unengaging instructional method is harder to understand, does not induce long-term learning, and decreases soldier motivation (Himmel, 1970; Keller, 1987).

"Army training and education methods evolve. The Army adopts better ways to foster learning, adapting how it trains units and develops leaders by employing innovative techniques, relevant to the learning requirements and environment" (Army Doctrine Publication 7-0, 2012, p. 1). The Army Learning Concept is an encompassing amalgamation of educational processes and models (epitomized in the Army Learning Model) that needs to be grasped in its many facets and applications both in techniques and environmental constructs for instructing students on all three learning domains: cognitive (knowledge/comprehension), affective (feeling/valuing), and psychomotor (kinesthetic/physical), also referred to as knowledge, skills, attitudes (KSAs) in academia.

THE ARMY LEARNING MODEL

The U.S. Army officially defines the Army Learning Model (ALM) in *Training and Doctrine Pamphlet 525-8-2, The U.S. Army Learning Concept for Training and Education, 2020-2040* as:

> The Army's adaptive, continuous learning model that is routinely improved to provide quality, relevant, and effective learning experiences through outcome-oriented instructional strategies that foster thinking, initiative, and provide operationally relevant context which extends learning beyond the

> learning institution in a career-long continuum of learning through the significantly expanded use of network technologies. (2017, p. 46)

In further explaining the Army Learning Model, the Army has gone on to state "The Army adapts and expands the current, learner-centric Army learning model and its continuously adaptive learning approach. The future career-long continuum integrates unit training, military education, self-development, and experience into a holistic learning approach" (TRADOC Pamphlet 525-8-6, 2018, p. 17). A key aspect of the Army Learning Model is that it is continually adapting based on the latest learning science; it is not fixed, except for its characteristics of ever-evolving, student-centered, critical thinking, experiential learning foundations.

As the Army Learning Model is student-centered, soldier-students' point of view, perceptions, and previous learning are to be taken into account and used to enhance the learning process as much as possible. The importance of students' previous learning/understanding implies that the Army Learning Model prescribes to the constructivism theory of general learning (Dewey, 1910; Piaget, 1985; Vygotsky, 1978). Harland, professor of learning theories, (2003) in researching Vygotsky, articulately describes constructivism "…there is always a strong relationship between what the learner already knows and can do, and what is to be learned. …each student will construct their own meaning based on an interaction between prior knowledge and current learning experiences" (p. 266). It is also important to note that the ALM's use of student-centered education also encompasses the notion of learning-centered. Although in some circles student-centered is viewed as specifically focusing on the individual student and learning-centered is viewed as focusing on the learning process itself, ALM incorporates both terms into student-centered, focusing on proven benefits of enhancing both the individual student's experience and the science of the learning process (Carlisle, 2016; Wright, 2011).

Another major aspect of the Army Learning Model, concerning its student-center ideal, is its focus on ensuring that students are specifically *motivated* to learn. John Keller's ARCS Model of Motivation (Attention, Relevancy, Confidence, Satisfaction) offers an easy to understand explanation of how to enhance student motivation to better enable learning (1987) – a deeper explanation of the ARCS model is provided later in the Motivation chapter of this book.

Although learning is often associated with either training (defined as field/unit task-based activity) or education (a specific learning institution or online Structured Self-

Development/Distributed Leader Courses), both training and education are vital elements that must occur within the operational force (TRADOC Pamphlet 525-8-2, 2017). Formal "book" learning and hands-on experiential learning can and should be done together in that effective use of the Army Learning Model incorporates both training and education. Instructors/facilitators must understand that training/education is conducted in three training domains: institutional (formal schooling), operational (learning on the job), and self-development (self-directed learning) (Army Doctrine Publication 7-0, 2012). ALM needs to be used to maximize the capability and results of each training domain.

*Note that although the Army Learning Model (ALM) mainly expresses the term "facilitator," the term used throughout this book will be "instructor/facilitator." ALM highlights "This learner-centered instructional approach encourages student participation and puts the instructor in the role of a facilitator" (TRADOC Pamphlet 525-8-2, 2017, p. 20). The term "instructor/facilitator" will be used in this document as a way to emphasize that even as a facilitator, that individual is also an instructor and must stand ready as a credible subject matter expert to address any question, maintain student focus, ensure that students fully comprehend the material, and grasp its importance/relevance (Kolbe & Kolbe, 2017).

HOW TO CONDUCT INSTRUCTION USING ALM

To conduct good, effective training that correlates with the Army Learning Model, four main overarching concepts must be fully understood: *motivation, interaction, critical thinking,* and *experiential learning*. Although an understanding of andragogy (adult learning) is also important, it will be addressed within these four main concepts as well as touched upon in a later section within this book. In the same way, the instructional underpinning of student-centered learning is expressed within these overarching concepts as well as mentioned specifically later within the book. These four main overarching concepts of *motivation, interaction, critical thinking*, and *experiential learning,* which are distinct yet overlap, are critical to maximize students'/soldiers' comprehension, learning, and long-term information retention. Each of these concepts will be described and an **Implementation Example** will be presented to enhance instructional practices.

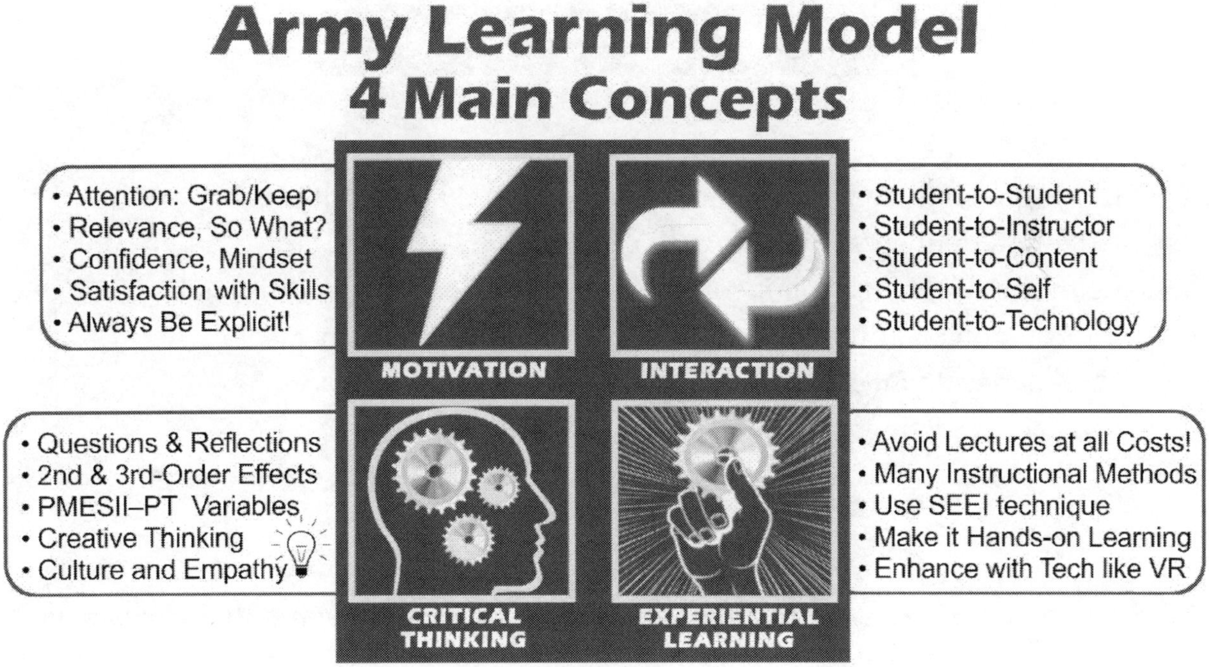

Four Main Concepts of the Army Learning Model

CHAPTER 2: MOTIVATION

- Introduction to motivation (defined/explained) along with its importance
- Attention defined/explained with examples
- Relevancy defined/explained with examples along with additional information on adult learns, Millennials, and Generation Z
- Confidence defined/explained with examples
- Satisfaction defined/explained with examples
- Implementation Example: Class Using the ARCS Model of Motivation

First and foremost, all educational personnel must understand that *motivation* needs to be maximized to help ensure students are fully primed to learn. This overarching concept of the Army Learning Model is described first because it is the most important. Many instructors, trainers, and even professors overlook this area and simply hope that students are intrinsically (self) motivated, and are in their class because they want to learn, sadly this is not always the case (Anders, 2014). Motivation needs to be fully understood and implemented to fully achieve the success that can be accomplished with the Army Learning Model.

The generally established definition of motivation deals with a desire or willingness to do something, or a reason for action/change in behavior. Keller (1987c) describes motivation as "…conditions that will stimulate people's desire to be interested and involved in their surroundings and to achieve their best" (p. 1). Within this document, motivation deals with the deliberate act of instilling a drive or reason for the student to want to learn (can also be the facilitation of awakening or directing students' own internal motivation). The concept of motivation is grandiose in that it overlaps many other important listed concepts and areas associated with the learning process. The value and importance of motivation is well described by noted researcher Wlodkowski (2008) "Motivated learners care more and concentrate better while they expend effort, and they are more cooperative. They are therefore more psychologically open to the learning material and better able to process information." Additionally, by specifically thinking about the motivational needs of the learner we are already creating a more student-centered learning environment.

Although there are many different theories and models addressing educational motivation, Keller's ARCS: Attention, Relevance, Confidence, Satisfaction model of motivation provides an easier to understand and very usable framework (Keller, 1987a). There is strong efficacy to use the ARCS model of motivation due to its utilization and endorsement by the U.S. Army and other portions of the military (Keller, 2008; Quinkert, Morrison, Fletcher, Moses, & Roberts, 2007; Reinerman-Jones, Lameier, Biddle, & Boyce, 2017). The ARCS model has four different sections that, when fully addressed, effectively improve student motivation. Although each section of the ARCS model could be used independently, the greatest effect (force multiplier) comes from ensuring that all aspects of motivation are addressed (synergistic effect) to increase the students' desire and persistence in learning.

Keller's ARCS Model of Motivation and How to Implement

ATTENTION: Grab and maintain students' focus through excitement, arousal, sensation-seeking, and curiosity.
Implementation: Lively/active discussions, simulations, role playing, use of stories, analogies, and appropriate humor.

RELEVANCY: "So What?" Why is this important personally now & in the future, show the value of learning.
Implementation: Explicitly discuss the relevancy, use reflective essay, show enthusiasm, bring in guest presenters.

CONFIDENCE: Students' belief that through their own efforts/persistence they will be able to learn and succeed.
Implementation: Clear learning goals via syllabus, grading rubrics, good instructional presence, & low stakes testing.

SATISFACTION: How content the student is with their educational accomplishment - overall experience.
Implementation: Provide timely feedback (both positive & negative), use of new skill, express learning value.

(Derived from Keller 1987 and Anders 2014)

ATTENTION

The first section of the ARCS model is *Attention,* which an instructor/facilitator must obtain from the student for learning to occur. If a student is not paying attention then the brain is not absorbing the instruction or educational experience (Keller, 1987a). Attention must be obtained and then **maintained** so that the student can continue to focus their attention on the instructional process. Excitement, arousal, sensation-seeking, and curiosity should be maximized in that the worst thing an instructor/facilitator can be is *boring* (Pekrun, Goetz, Frenzel, Barchfeld, & Perry, 2011). A sense of boredom is experienced by the student if their instructor uses poor interaction, ineffective instructional techniques, and fails to address the topic's relevancy. Keller recommends using instructional techniques that require learners to actively participate, such as through simulations, role-playing, discussions, and games. Additionally, using creative stories, situations, appropriate humor, and content analogies, along with regular problem-solving activities keep students engaged, interested, active (experiential learning), and maintains their attention (Keller, 1987a). What I tell instructors/professors is not to think of their class sessions as simply a class to attend, instead, it should be viewed as a special event; make it memorable, something that your students will be thinking and talking about even after the class-event is over!

RELEVANCY

The second section of the ARCS model of motivation is *Relevancy*, which answers the "so what" question that many students contemplate when having to learn new content. I believe that this is the most important/powerful aspect of motivation: relevancy.

All instructors/facilitators should express the subject matter's relevance as part of the first thing they do before instruction even begins (even before stating tasks, conditions, and standards). Students need to fully understand why the content to be learned is important now as well as in the future and why it is specifically relevant to them (personalization). Why is the content useful? These questions should be answered explicitly by the instructor as well as by the students themselves. Keller also recommends that instructors/facilitators "ask learners to relate the instruction to their own future goals" (Keller, 1987a, p. 4). This could be done in several ways such as an in-depth discussion (as a class or in groups) or an assignment such as a reflective essay.

Relevancy of content to be learned can also be enhanced by how important/valuable the content is perceived to be by the student. Instructors/facilitators must model enthusiasm for the subject that is being instructed; accomplished by their comments about the subject, interaction techniques, instructional presence, and displayed levels of energy for the learning material (Anders, 2017). Never express negative feelings about the content to be learned. An off the cuff comment like, "I know this is boring and we've had this class before, but we just got to get through this," is the same as telling students this information isn't important, don't bother paying attention. Enhancing the value of the learning material can also be accomplished by taking the time and effort to bring in outside guest speakers that also express high enthusiasm for the educational content/subject matter (Keller, 1987a). Building excitement about the subject matter helps to build relevancy and create an emotional connection with the content being learned which also enhances motivation and learning (Hascher, 2010).

Relevancy also plays a key role in helping/nurturing students develop and commit to educational goals for success, "Teachers and parents often assume that students share a commitment to academic goals, whereas the reality is that developing this shared commitment needs to be nurtured and built" (Hattie & Timperley, 1981, p. 89). Relevancy enhances commitment by building up students' intrinsic motivation (inner desire/interest to accomplish/experience/learn). Intrinsic motivation is powerful in that it helps students become lifelong learners. The power of *Relevancy* to help students understand the importance and commit to educational goals applies to all types of learners.

The Adult Learner

Although many may think that relevancy (answering the "so what) is predominantly important with millennials/generation Z (younger students), it is also a highly important aspect with regards to adult learning, yet who exactly is the adult learner? Even with a Ph.D. in the specific field of Adult Learning, I have to tell you that the specific answer is that it very much *depends*. Some might try to define an adult learner as anyone who is legally an adult (over 18) or who has finished general formal education (over 22). Others might define adult learners as those with competing interests due to adult responsibilities such as jobs and family (Bourdeaux & Schocnack, 2016). Although these are generally agreed upon adult learner characteristics, this is still an oversimplification in identifying the Adult Learner.

Malcom Knowles (famous adult learning researcher) described adult learning as existing more on a continuum of learning complexity and of learner ability/cognitive capability (1973). In describing an adult learner Knowles stated: "…one who has arrived at a self-concept of being responsible for one's own life, of being self-directing" (1984, p. 9). Self-directing meaning able to do more with less supervision and being somewhat more self-motivating. Brookfield (1995), another major adult learning researcher, describes adult learning as pertaining more to the level and quality of the learner's previous experiences, the nature of the learning task, and subject matter involved. This introduces another important complexity to the aspect of adult learning. If I already know something about the subject or learning task (I have some experience) then I would generally be more of an adult learner; if I have less familiarity with the subject (I have little to no experience) then I would be much less of an adult learner. For the specific purposes of this document, the *adult learner* is defined as anyone old enough to be in the military and has a prior knowledge base - experience in what is being learned. This individual could be more of an adult learner (continuum) based on a combination of rank, years of service, and level/quality of experience (and understanding) in the subject matter being learned.

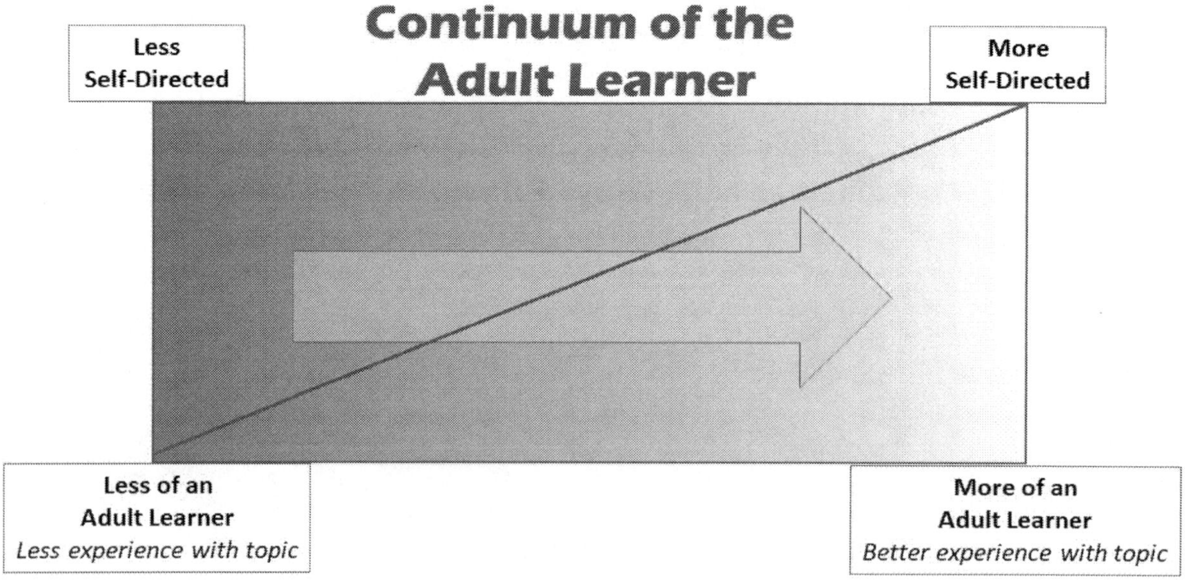

There are six generalized core principles of adult learners that help to understand their thought processes and motivations (Knowles, Holton, & Swanson, 2005).

1) Learner's Need to Know: adult learners tend to want to know more of the why, what, and how about a subject. A good way to address this is by properly informing students about the specific goals and topics of the course (should be fully addressed within the course syllabus). Part of this core principle also entails adult learners wanting to participate in the planning process for their own learning. Allow student input and provide options whenever possible (this also helps students to have a greater sense of ownership for their own learning).

2) Self Concept of the Learner: adult learners have a general desire to be autonomous and self-directing. When possible allow choices/options with regards to assignments or specific subjects and always encourage the development of self-directed life-long learning skills.

3) Prior Experience of the Learner: instructors/facilitators must recognize and understand the previous experiences (both type and quality) of adult learners to use as a resource and to build upon (enhancing thinking processes).

4) Readiness to Learn: adult learners need to see relevancy/usefulness of the learning material to their own, current, real-life (express relevancy as previously described).

Additionally, adult learners may need changing levels of assistance, learning scaffolding in order to succeed (be cognizant of student struggles and provide needed support).

5) Orientation to Learning: adult learners generally have a problem-solving learning orientation and they prefer experiential (hands-on) learning. Use action-orientated educational methodologies such as scenarios, case-studies, and frame instructions as problems to figure out together as much as possible/logical.

6) Motivation to Learn: adult learners tend to be more intrinsically motivated; current personal relevancy is extremely important as to why they are learning the material (be sure to explicitly answer the "why this is important" question for them).

CORE ADULT LEARNING PRINCIPLES

Learner's Need to Know

- Need course goals/topics: answer why, what, & how
- Want engagement in collaborative planning process for own learning:provide options when possible

Self-Concept of the Learner

- General desire to be autonomous & self-directing
- Encourage life-long learning skills whenever possible

Prior Experience of the Learner

- Recognize and understand previous experience type & quality to use as a resource and to build upon

Readiness to Learn

- Need to see relevancy/usefullness to real life (now)
- May need changing levles of assistance/scaffolding

Orientation to Learning

- Prefer problem-solving learning orientation
- Prefer experiential (hands-on) learning

Motivation to Learn

- More intrinsic, internal satisfaction
- Importance of personal relevancy now

Dirived from (Knowles, Holton, & Swanson, 2005)
- incorporated with additional explaination (Anders)

Knowles and his team also specifically expressed "Adults tend to be more motivated toward learning that helps them solve problems in their lives or results in internal payoffs" (Knowles, Holton, & Swanson, 2005, p. 199). Although most adult learners are generally internally motivated (intrinsic instead of extrinsic, meaning the knowledge/completion is more important to them than a grade or recognition), adults most often want knowledge that is currently *relevant* and can be immediately used for their job or mission (Merriam, 2001). Instructors/facilitators need to be explicit in helping all students fully realize the content's relevancy as well as understanding and addressing the other presented adult learning core principles.

Millennials and Generation Z

Despite a lot of hype and term creation, there really isn't that much difference between adult learners, millennials (born 1989 to 1999), Generation Z (born 2000 to 2019), or most students for that matter. Trent (2019) provides a good summary of current research dealing with motiving and educating millennials but many of the items noted are good items to apply to all instruction. Items such as wanting to know relevancy, having more choice in assignments/learning, having more support (learning scaffolding) are important to millennials but are also important aspects of learning for all. Yes, you do need to think about your students and their perspectives. Younger generations have a much deeper connection with technology (it's always been a part of their lives); it is a major emotional event to forbid/take away their cellphones in class (Boucher, 2019; Konok, Gigler, Bereczky, Miklosi, 2016; Mihailidis, 2013). Instead, look for ways where students can use their own personal technology to enhance learning (O'Bannon, & Thomas, 2015; Wood, Mirza, Shaw, 2018). Another powerful technique is to teach students how to be mindful (active conscious awareness of one's experienced phenomena) so as to avoid nomophobia (fear of not being able to use one's phone or of not being reachable through one's phone) by reflecting on their actions and understanding when it is appropriate (rational) and inappropriate (irrational) to use their cellphones (Arpaci, Baloglu, Kesici, 2019; Miller & Brannon, 2017).

Millennials and Generation Z students are accustomed to higher-end more appealing instruction with fun and even game-like interactivity at times (Trent, 2019). A great line is quoted by Trent (2019) in referencing Rickes' (2016) research article dealing with Generation Z "…Gen Zers will continue to prefer practical and hands-on learning given their desire for meaningful experiences. This predisposition will continue to raise the bar on active learning classrooms and pedagogy" (para. 60). Does that information sound familiar? It should because that is exactly what the Army Learning Model is all about, that is what this entire book is all about. ALM seeks to create powerful, meaningful learning experiences through practical hands-on learning to achieve greater understanding and long-term retention.

CONFIDENCE

The third part of Keller's ARCS model of motivation deals with the *confidence* that students have in their ability to actually learn the content being presented. This ties in directly with students having the right mindset (growth mindset as opposed to a fixed mindset) and believing that it is through their own efforts/persistence that they will be able to learn the material (Dweck, 1986). If a student doesn't personally believe that they can ever accomplish, do, or learn, what the instructor is teaching/demonstrating (having a fixed mindset), then the student will lose their motivation to learn; students must remain *confident* in their ability to comprehend and learn (Keller, 1987a).

Recommendations for enhancing student *confidence* are: provide clear, appealing learning goals as well as the specific criteria that will be used for grading (examples include a well-written *syllabus* and use of easy to understand grading *rubrics*). Materials should be presented in logical difficulty progressions so that students can adequately address the learning task (step-by-step, simplistically at first and then under harder, more realistic conditions). Helping students set realistic goals and helping them to organize and schedule their work and studying will also assist students' confidence. Encouraging students and attributing their success to their effort, helps students develop a greater sense of their own capabilities. Additionally, providing students an opportunity to engage in *low-stakes testing* (testing without large impacts on students' grades) helps the learner gain skill mastery thereby increasing confidence (Keller, 1987a). Using good verbal and non-verbal forms of communication through enhanced instructional presence will also help students feel like a contributing member of the learning community, help them be able to follow along with the instruction, and help them be confident in their own ability to learn (Anders, 2017; Bourdeaux & Schoenack, 2016; Keller, 1987b).

Confidence with regards to adult learners should be contemplated on multiple levels due to various possible characteristics of this type of learner. An adult learner may lack confidence dealing with the presented courseware due to extensive time passing since the last time they were exposed to the topic, unfamiliarity with the technology being used for the instruction, or unfamiliarity in the present learning environment (Knowles,

Holton, & Swanson, 2005; Merriam, 2001). Additional encouragement and educational resources, such as supplementary, optional learning modules would help to address adult learner concerns by allowing them to practice and/or catch-up on any areas lacking in their educational framework (also known as *learning scaffolding*). Instructors/facilitators at all levels should be mindful in developing students' confidence and ability by helping them become more independent, self-directed, life-long type learners to be able to overcome any educational deficiency in the future.

SATISFACTION

The fourth and final section of the ARCS model of motivation pertains to how *satisfied* or content a student is with their educational accomplishment. "To have a continuing desire to learn, the student must have a sense of Satisfaction with the process or results of the learning experience" (Keller, 1987b, p. 2). Student satisfaction is greatly enhanced when they receive proper feedback (satisfaction with the process) and can accomplish success (satisfaction with the results).

Although receiving praise is an important component (and psychological enhancer: extrinsic motivation), proper feedback is made up of much more. Feedback must be given as immediately as possible and needs to be personally relevant with unambiguous information that will help that specific student (O'Donovan, Rust, & Price, 2015). Although giving a compliment such as "great job" is a beneficial motivational feedback statement, it shouldn't be the only type of feedback provided. Timely feedback covering what the student did wrong (with specifics on how to improve) as well as what the student did right (both positive and negative feedback) should be given during each feedback session. Both areas (what was done right and wrong) are important for students' comprehension and development so that they can continue to do what was right and work on improving what was done wrong. When giving feedback it is important to not only address the specific task but to provide feedback regarding the general process for generalizability as well (Hattie & Timperley, 1981; O'Donovan, Rust, & Price, 2015). Students should also receive feedback on their thinking process: *how did they come up with that solution, were they viewing the problem correctly,* in order to help them gain more *metacognition* (ability to think about their thinking) and develop critical thinking skills (Wieman, 2007; Lovelace and College, 2014).

Excessive compliments should be avoided but feedback should never be completely negative; properly phrased and balanced constructive criticism and praise is the goal (Nilson, 2010). Multiple research has found that properly facilitated feedback provides students with an enhanced educational experience which translates to greater satisfaction (Gerdera, Williams, & Wright, 2015; McIsaac, Blocher, Mahes, & Vrasidas, 2006). Additionally, feedback goes both ways.

Receiving feedback from students provides facilitators/instructors with valuable insights dealing with the educational content and process. Reynolds, Roberts, and Hauck (2017), in researching the use of Keller's ARCS model of motivation, identified additional benefits to receiving student feedback "Giving students opportunities to provide feedback on sessions makes them feel valued and gives them a sense of control (intrinsic reinforcement)" (p. 18). In addition to student-to-instructor feedback, peer (student-to-student) feedback has also shown to improve satisfaction and improve the learning process. Nilson's book "*Teaching at its Best: A Research-Based Resource for College Instructors*" addresses the benefits of peer feedback and claims that it helps learners develop communication, collaboration, critical thinking, and lifelong learning skills (2010). Student-to-instructor and student-to-student feedback also addresses adult learning in providing these learners with a means for expression, use of their personal experience, and a sense of respect (Knowles, Holton, & Swanson, 2005; Merriam, 2001).

Peer feedback and student-to-instructor feedback can be accomplished through any combination of *After Actions Reviews* (AARs), in-class or online discussions, end of class/course critiques, or even use of in-class clickers (digital polling tool used to answer questions and provide class feedback).

Another very important component of *Satisfaction* is achieved by providing students with an opportunity to succeed by using their new skill (what was learned) in a realistic setting (Keller, 1987b). Realism is enhanced by adding sights, sounds, smells, and stressors that more closely match what a real situation would be like pertaining to the learned skill. This can also relate to low-stakes testing but specifically deals with helping the student show to themselves and others that they can accomplish and use what was learned effectively (this also falls in line with experiential learning). This opportunity can take many forms such as in-class demonstrations, going out to the range or special real/realistic environment, creating video/computer simulations, using scenarios, etc. (Nilson, 2010). Creativity should be used to (once the student is ready) provide an opportunity/event where the student can use the learned skill and succeed, thereby instilling a sense of accomplishment and self-worth, equaling satisfaction.

The motivational aspect of *Satisfaction* also pertains to students' overall experience with the educational endeavor. How worthwhile was the process and was it worth the effort/time/expense, are all contemplated by students as they determine their overall satisfaction. In light of this understanding, instructors/facilitators should be explicit in helping students realize the value of what they have learned and the learning event (Nilson, 2010). Although this overlaps with the aspect of *Relevancy,* students should be reminded why this information is important, how their own hard effort paid off, and how their newly learned skill will benefit them now, as well as in the near and far future (how is knowing "this" a valuable usable skill), (Keller, 1987a).

By understanding what motivation is and how to apply it to education, instructors/facilitators can greatly improve students' learning outcomes. Keller's ARCS model of motivation provides an easy to understand breakdown of motivation into the four different sections of Attention, Relevancy, Confidence, and Satisfaction. View the following infographic for a summary description and implementation guide.

Keller's ARCS Model of Motivation and How to Implement

ATTENTION: Grab and maintain students' focus through excitement, arousal, sensation-seeking, and curiosity.
Implementation: Lively/active discussions, simulations, role playing, use of stories, analogies, and appropriate humor.

RELEVANCY: "So What?" Why is this important personally now & in the future, show the value of learning.
Implementation: Explicitly discuss the relevancy, use reflective essay, show enthusiasm, bring in guest presenters.

CONFIDENCE: Students' belief that through their own efforts/persistence they will be able to learn and succeed.
Implementation: Clear learning goals via syllabus, grading rubrics, good instructional presence, & low stakes testing.

SATISFACTION: How content the student is with their educational accomplishment - overall experience.
Implementation: Provide timely feedback (both positive & negative), use of new skill, express learning value.

(Derived from Keller 1987 and Anders 2014)

IMPLEMENTATION EXAMPLE: Class Using the ARCS Model of Motivation

In this scenario, the example class deals with instructing student-soldiers on how to properly emplace and use an Army tourniquet.

"Welcome everyone. If you look in our syllabus you will see that today we are focusing on how to properly put on a tourniquet, not just on yourself, but on someone else as well. This is important because with this valuable knowledge you will be able to save someone's life and it could save your own life as well. Now it's one thing for me to tell you about it, but I think it will be even more impactful if you hear it from someone with a first-person account. I would like to introduce you to our special guest presenter that we have with us here today. Please welcome combat medic SGT Smith."

SGT Smith would then proceed to tell his impactful and very engaging story about how, while on his last deployment, an improvised explosive device (IED) had gone off and torn apart a fellow soldier's arm. SGT Smith would describe, in detail, how that soldier was bleeding to death but he (SGT Smith) was able to get to him, get an Army tourniquet on the soldier, and save his life. Yet the story wouldn't end there. SGT Smith would explain that after they had medically evacuated (MEDEVAC) the soldier, they continued with their patrol until SGT Smith accidentally stepped on an IED himself. The explosion blew off his lower left leg causing him to begin to bleed to death as well. None of the other soldiers could reach him because a firefight had ensued. Luckily, SGT Smith was still conscious and was able to fight through the pain and the ringing in his ears to emplace a tourniquet on himself to stop the bleeding. At that point SGT Smith would lift his pant leg, revealing a sophisticated prosthetic leg.

"I owe my friend's life as well as my own life to the information I learned in a classroom just like this. Please pay attention and remember what you learn about properly emplacing an Army tourniquet."

In this example, using a guest speaker like this towards the beginning of class has functioned to address two ARCS model of motivation components: *Attention* and *Relevancy*. A powerful, engaging, and real-life story like that of SGT Smith would grab and maintain students' *Attention* as well as provide important personal *Relevancy*. The fact that SGT Smith's personal story is specifically about Army tourniquets would also help to focus the student-soldiers on the task at hand (how to properly emplace an Army tourniquet).

"Thank you, SGT Smith, for that riveting account about your real-life use of the Army tourniquet. We are all very thankful that you are here today to share this important and extremely *relevant* lesson with us. "

"I'm now handing out the Army Combat Application Tourniquet™, or CAT for short. These are just like the ones that SGT Smith used. Before we leave here today, each one of you will be able to emplace the CAT on yourself and on another soldier within 60 seconds and in accordance to the tasking steps as outlined in the read-ahead homework found in your Solider Handbook of Common Tasks (tasking number: 081-COM-0048)."

"First, I'm going to show you how to put it on yourself at full combat speed, and then I'll show you again, step-by-step and fully explain each stage of the process. When I show you each step, you will be following along and emplacing the tourniquet on yourself to get a first-hand experience of the process. Ask me questions at any time if you need further clarification. We'll have plenty of time to practice and we'll play a game and have a little fun competition later to see who can put it on themselves and others the fastest while still doing it correctly. All of you will be able to do this, just pay attention and follow along, remember how important this is for you and your fellow soldiers. After we are done we will discuss and have an After Actions Review."

During this part of the example scenario notice how the *Confidence* aspect of ARCS is addressed through the use of the instructional technique which in this case is very hands-on (experiential). Student-soldiers are able to physically touch the tourniquet and follow along as the instructor explains the process step-by-step; this would greatly lower students' anxiety about the event while increasing their *confidence* about how the tourniquet application works. Also, notice the instructional presence displayed by the instructor/facilitator. The instructor/facilitator thanks the guest presenter, tells the students to feel free to ask questions at any time, clearly expresses the task, and uses phrases such as "we'll have plenty of time," and "we'll have a little *fun* competition." All of these things work to make the instructor that much more approachable, part of the group, and allows for low-stakes testing which helps the students and increases their *confidence*. Also notice how the instructor offers very specific information on what will occur and what the evaluation criteria "within 60 seconds" will be (this ties in with having a clear syllabus and easy to understand grading rubrics).

"Everyone did a great job in the competition. I worked with each of you one on one, gave you feedback, and then together we were able to get everyone to be able to do it in 60 seconds or less! That's awesome. Special kudos to our overall winner SPC Anderson with a time of 26 seconds; that is amazing. Also, thanks to everyone who turned in a class evaluation/comment card, we will definitely look into the great feedback you provided about the class and instruction."

The next day:

"Today we are going to push our capabilities even more. I brought all of you outside today, wearing your helmet, protective vest, and gloves so that we can practice then test emplacing a tourniquet in a more realistic way. We are going to take what we learned in the classroom and apply it in a new way."

"Here is what we are going to do; you're going to run as a two-man battle team, as quickly as possible, to a down soldier 50 meters away, it's a mannequin that weighs 150 pounds. Talk to your partner and work together to move the downed soldier properly and efficiently. In this scenario there is shooting in the area, you will hear sound effects from these outdoor speakers. The area is not secure, but others are providing you with cover so you can reach the soldier. You will have to move the soldier back to this location, as quickly as possible, where it is fully secure. Your heart rate will be pounding by this point, just like in real life. As soon as you get the soldier-mannequin here, evaluate and then properly emplace a tourniquet within 60 seconds, just like we practiced."

"Let's put what we've learned into real practice. Show me and your fellow students how well you are now able to emplace a tourniquet at full combat speed.

Remember, it might just save a life, to include your own, someday. After we are all done, we will have an After Actions Review. Alright, let's do it!"

The final aspect of Keller's ARCS model of motivation, *Satisfaction*, is addressed in a multiple of ways. Feedback is a major theme here, with two-way communication; feedback to the student and feedback from the student to the instructor (class evaluation/comment card). Another powerful implementation that affected *satisfaction* in this example was when the instructor/facilitator provided the students with an opportunity to succeed by using their new skill (emplacing a tourniquet) in a new way, in a realistic setting (outside, wearing combat gear, with sound effects, and after having to move a down soldier-mannequin). After completing this more realistic task, students will have shown themselves and others their newly enhanced capabilities resulting in contentment with the educational endeavor and an overall high level of *Satisfaction* with the experience.

From this example, it is important to note that the instructor was able to address each aspect of Keller's ARCS model of motivation. Notice how the language used was very explicit during the motivational efforts at the beginning, with the content explanation, and with the practical exercise/assessment instruction towards the end. The instructor/facilitator used keywords like "relevant," "we," as well as made sure that all understood what was happening "you will be able to emplace the CAT on yourself and on another soldier within 60 seconds," "all of you will be able to do this." Additionally, note how the class became hands-on as early as possible, allowing students to touch a tourniquet and begin to practice early on. These words/phrases and techniques are specifically used to help address all aspects of (ARCS) motivation, being as direct as possible and avoiding ambiguity or hoping for students to grasp implied importance.

SUMMARY: Motivation

Motivation, presented as the most important component of ALM, must be the first thing to be addressed in the classroom to ensure students pay attention, focus on the instruction, and persist in learning the material. This chapter focused on the overall aspect of motivation and provided Keller's ARCS Model of motivation as the main way to comprehend this issue. Information was presented dealing with all aspects of motivation: Attention, Relevance, Confidence, and Satisfaction. Special emphasis was given to relevancy in that this special motivational component comes up again and again within students' psyche and educational mentality. Adult Learners, Generation Z, and Millennial learners were discussed along with their differences, similarities, and expectations. An example class was provided to highlight how the ALM aspects of motivation should be integrated into instruction to maximize understanding and learning.

CHAPTER 3: INTERACTION

- Introduction to *interaction* (defined/explained) along with its importance
- Student-to-instructor interactions defined/explained with examples
- Importance of instructional presence and cultural considerations
- Student-to-student interactions defined/explained with examples along with information dealing with group work (techniques and assessments)
- Student-to-instructional content interactions defined explained with learning objective information covering cognitive, affective, & psychomotor domains
- Emotional Connection and fun to create impactful learning experiences
- Movement/Exercise to enhance learning and student cheating issues
- Student-to-self: reflection and self motivation
- Student-to-technology: slideshows, video, instructional technology

The Army Learning Model overarching concept of *interaction* can best be defined as students' *actions* or *communications* with the instructional community, environment, and/or learning content. "Army institutions create and sustain a learner-centric environment by focusing on the dynamic **interaction** between faculty, students, and relevant, outcomes-based programs of instruction or curricula" (Training and Doctrine Pamphlet 525-8-2, 2017, p. 17). The importance and benefits of different types of interactions have been highlighted many times by multiple research for both face-to-face and online learning (Bernard, Abrami, Borokhovski, Wade, Tamim, Surkes, & Bethel, 2009; Jacobs, Renandya, & Power, 2016; Moore & Thompson, 1997; Nilson, 2016). As an example, in a meta-analysis of 1,034 educational research articles (Bernard et al., 2009), it was revealed that increasing the interaction component within online learning courses "...positively affects student learning" (p. 1264). In a similar matter, but with regards to face-to-face instruction, noted anthropology professor Michael Wesch explains that true learning and student transformation comes through interaction on all levels (2017). In addition to the three main established types of educational interaction famously highlighted by Moore (1989) consisting of *student-to-instructor*, *student-to-student*, and *student-to-instructional content*, this book will also highlight the importance of understanding the educational interaction aspects of *student-to-self* and *student-to-technology*.

STUDENT-TO-INSTRUCTOR INTERACTION

Although this interaction type is student-to-instructor (or vice versa: instructor-to-student), this also applies to student-to- "facilitator" (instructor/facilitator). Student-to-instructor interactions deals specifically with all the different communications, actions, and exchanges that occur between the instructor/facilitator of the course and the student(s). This would involve interactions between the student and the instructor both during class and out of class given that even face-to-face courses now tend to have asynchronous (out of the physical classroom/time) components. These interactions are comprised of verbal communication (how an instructor speaks to the class as well as one-on-one), non-verbal communication (an instructor's body language and how he/she carries themselves in the class), as well as written communication such as through assignment/test feedback whether on a paper document or online through an instant messenger, graded digital assignment, or email exchanges.

How an instructor ***verbally*** interacts with students is important due to aspects of comprehension and motivation. Both comprehension and motivation are enhanced when instructors/facilitators master the 6 main verbal communication characteristics: 1) projection, 2) pitch, 3) pace, 4) pauses, 5) pronunciation, and 6) avoidance of verbal distracters (Army Basic Instructor Courseware, 2012; Bambaeeroo & Shokrpour, 2017).

Effective *projection* requires the instructor/facilitator to understand how loudly he/she needs to speak not only for the students to hear the instruction, but to set the tone of the educational environment (an advanced course where more discussion will take place versus a basic training situation where soldiers are learning more basic skills, such as Army Basic Training).

Pitch is also very important in that a monotone voice without variations in pitch can result in a dull voice which, as described in the section dealing with motivation, can cause a student to lose interest and not focus/learn the instruction being presented.

Pace is significant in that if an instructor/facilitator talks too quickly many students may not be able to keep up and fully understand, if an instructor/facilitator talks too slowly, many students will get bored and lose interest.

Effective use of *pauses* greatly enhance instruction by…adding emphasis/dramatic effect, as well as focusing the students' attention, and giving students time to think about the information just presented.

Proper *pronunciation* is necessary to ensure that students clearly receive the information being presented and to maintain instructor credibility (professional instructors/facilitators will be able to properly pronounce all words needed within their block of instruction). The final aspect of *avoiding verbal distractors* deals with avoiding any type of poor verbal communication (slurred speech, stuttering, overusing a specific word: "like, so, ok, uhm, you know," or excessively clearing the throat or making any other type of sounds) that might distract a student from the learning content/process. Other things to consider are verbal tone and word choice (appropriate for the audience); students should be treated with respect and not talked down to in order to avoid student resentment and having them shut down (avoiding instruction).

How an instructor interacts with students in a ***non-verbal*** way is also important, again due to the aspects of comprehension and motivation. Understanding the intended message and obtaining and maintaining student motivation are enhanced when instructors/facilitators master the 6 main nonverbal communication characteristics: 1) general appearance, 2) posture, 3) body movement and distance, 4) eye contact, 5) gestures, and 6) avoidance of non-verbal distracters (Army Basic Instructor Courseware, 2012; Bambaeeroo & Shokrpour, 2017).

Although some might view one's *general appearance* as an aspect of vanity, it is important to always look at things from students' points of view. Instructors'/facilitators' appearance is viewed as an important aspect of their professionalism and credibility; as such one should be an exemplar of proper hygiene, well organized, be well-groomed, physically fit, and be in the right uniform or professional attire (Behling & Williams, 1991).

Posture is key in that a good upright stance exudes confidence and enthusiasm for the lesson material which translates into increased student motivation. Avoid hunched over shoulders and presenting an "I don't care" body language message. Students are greatly affected (either consciously or subconsciously) by what they see; they should believe that you truly care and are interested in the instructional material just by looking at you (Gurney, Howlett, Pine, Tracey, & Moggridge, 2017).

Body movement and distance is a powerful tool that can be used to encourage discussion (by moving closer or even sitting down), to emphasize an important point, to transition to a new point, or to help address negative student behavior (by moving closer to them).

Eye contact is an important communication tool that enhances one's credibility (people who don't make eye contact tend to be viewed as hiding something and unreliable), increases motivation, and aids the instructor/facilitator with a greater opportunity to observe the students, to gauge their comprehension (additionally to identify any individuals causing distractions).

Gestures are another important tool in that when used properly they help to explain the content being presented, can add emphasis, and can be used to encourage or express understanding during active listening (head nodding indicating comprehension). The aspect of avoiding *non-verbal distractors* is accomplished by being aware of items such as excessive pacing, fidgety hand movements (twirling thumbs, playing with keys or change in pockets, etc.), or other distracting movements. A great way to identify any of these distractors is to video record oneself while instructing. Additionally, instructors/facilitators should not forget to **smile** when appropriate to enhance approachability and lower students' anxiety (Van Kleef, van den Berg, & Heerdink, 2015). I have tried to smile (especially on the first day of class) as much as appropriately possible and have received many end-of-course-critique comments from students about it aiding in my approachability and in lowering their initial nervousness about the course.

Another way that students and instructors/facilitators commonly interact is through written means (whether on printed documents or online). Effective written communication can be comprehended by the reader in one quick reading (concise), is properly organized, and is free of errors (Army Regulation 25-50, 2013). This is accomplished by reviewing all correspondence (read it twice) before it is given to the recipient, putting the main point of the correspondence (recommendation, conclusion, or reason for writing) at the beginning (bottom line up front) and by using *active voice* (Department of the Army Pamphlet 600–67, 1986).

Active voice consists of creating shorter sentences by putting emphasis on the actor of the sentence by placing the main subject before the verb; it is made up of the standard English sentence order of subject-verb-object (example: You will have up to 60 seconds in which to emplace the Combat Application Tourniquet™). This type of effective communication is beneficial in that it increases the speed of communication and assists the reader in better understanding the main point of the message (Army Regulation 25-50, 2013). Be sure to also be very careful with humor and sarcasm within written text since one's tone is not usually fully understood via written form (this can lead to misunderstanding).

Yet effective verbal, nonverbal, and written communications are not enough to achieve the most beneficial student-centered, student-to-instructor interaction. During student-to-instructor interactions, the instructor/facilitator must maintain a mindset of not just giving information so the student can pass a test, but instead of helping the student be part of a community of inquiry where the student is positively transformed and knowledge is effectively retained for critical diverse implementation when needed. This improved instructor mindset and enactment is accomplished through enhanced *instructional presence* (Anders, 2017).

Instructional Presence

Effective *Instructional Presence* is the culminating result of the active use of the three components of the Community of Inquiry Model: *Social Presence, Cognitive Presence,* and *Teaching Presence* (Garrison, Anderson, & Archer, 1999). As a life-long learner (both in the civilian world and in the Army), having taken thousands of online and face-to-face courses, I would best describe Instructional Presence as:

> …the ability of a teacher to make a student feel like a real learning member of an instructional community and not "just a number," that is not simply an unknown student just going through the motions of taking the class. (Anders, 2017, p. 7)

Social Presence

The first main component of instructional presence is *Social Presence,* which refers to the idea of social respect by valuing everyone's opinion, providing timely feedback, and humanizing the learning environment/community. This is done by conducting instructor and student introductions (showing one another that we are real human beings) as well as through properly facilitated discussion on the learning material. These processes help students achieve a greater sense of connectedness within the instructor and one another, enhancing the overall educational experience. Another way to aid in the humanization of the classroom can be accomplished through the proper use of humor (never derogatory and always appropriate for the audience/environment). Correlating with the proper use of humor, is the establishment of an educational environment where the students feel free and safe to expresses their thoughts and emotions so that the student is able to fully reflect on what was learned and help to establish an emotional connection to the content which aids in long term learning (Doyal, 2011).

As instructors/facilitators, we must readily stay on topic, but also be flexible enough to let students speak and fully discuss a topic. Properly facilitating a discussion doesn't mean giving all the answers or sitting back and doing nothing. It means guiding the discussion just enough so that all feel welcomed, have a sense of trust, can participate-communicate freely, can arrive at the right answers, and are willing to help listen/learn and help out one another (Anders, 2017).

Cognitive Presence

Cognitive Presence deals with ensuring that students go through all parts of a learning process described by Garrison, Anderson, and Archer as the Practical Inquiry Model (1999):

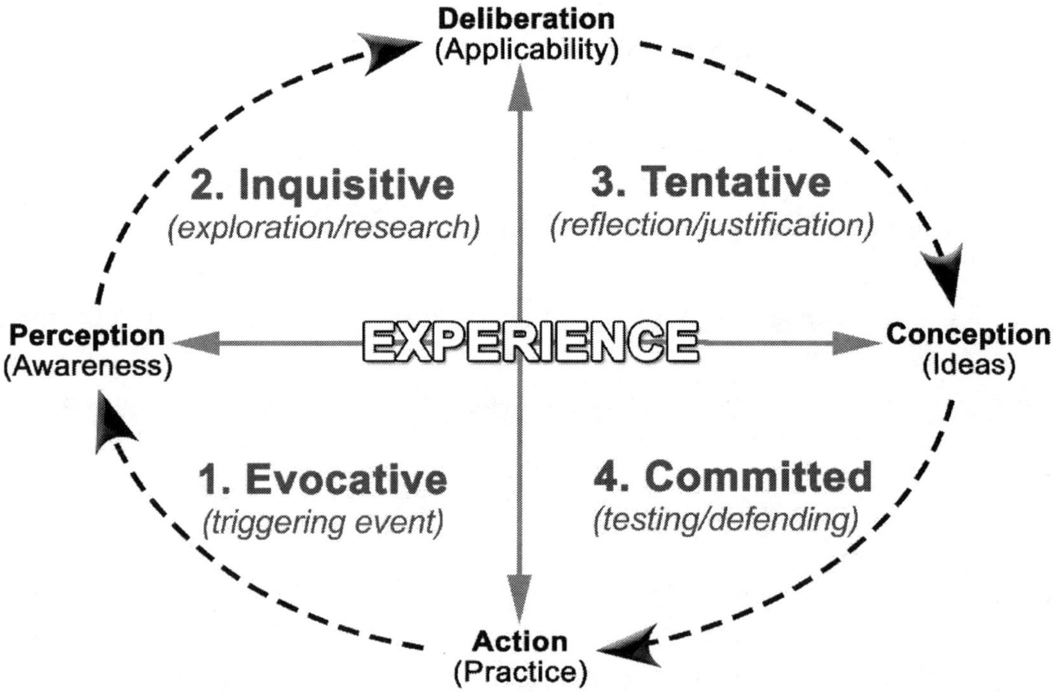

Representation of Practical Inquiry Model by (Garrison, Anderson, & Archer, 1999)

Phase One: Evocative, is experiencing an actual event and having an awareness of it occurring. Phase Two: is Inquisitive which deals with the student exploring and researching what occurred during the event and starting to contemplate its applicability. Phase Three: Tentative involves active reflection and justifying the learned event's usefulness and incorporating it into previously gaining knowledge. Phase Four: Committed, is the final phase and deals with action, applying (testing/experimenting/using) the knowledge learned in the "real world."

This entire learning process involves critical thinking and was originally described by Dewey (1910), then further developed by Kolb (1984) and Garrison, Anderson, and Archer (1999). Kolb's Experiential Learning Model is what is most often cited in most Army writings and is an important aspect of the Army Learning Model which uses a slightly modified version referred to as the Army Experiential Learning Model (TRADOC Pamphlet 350-70-7, 2018).

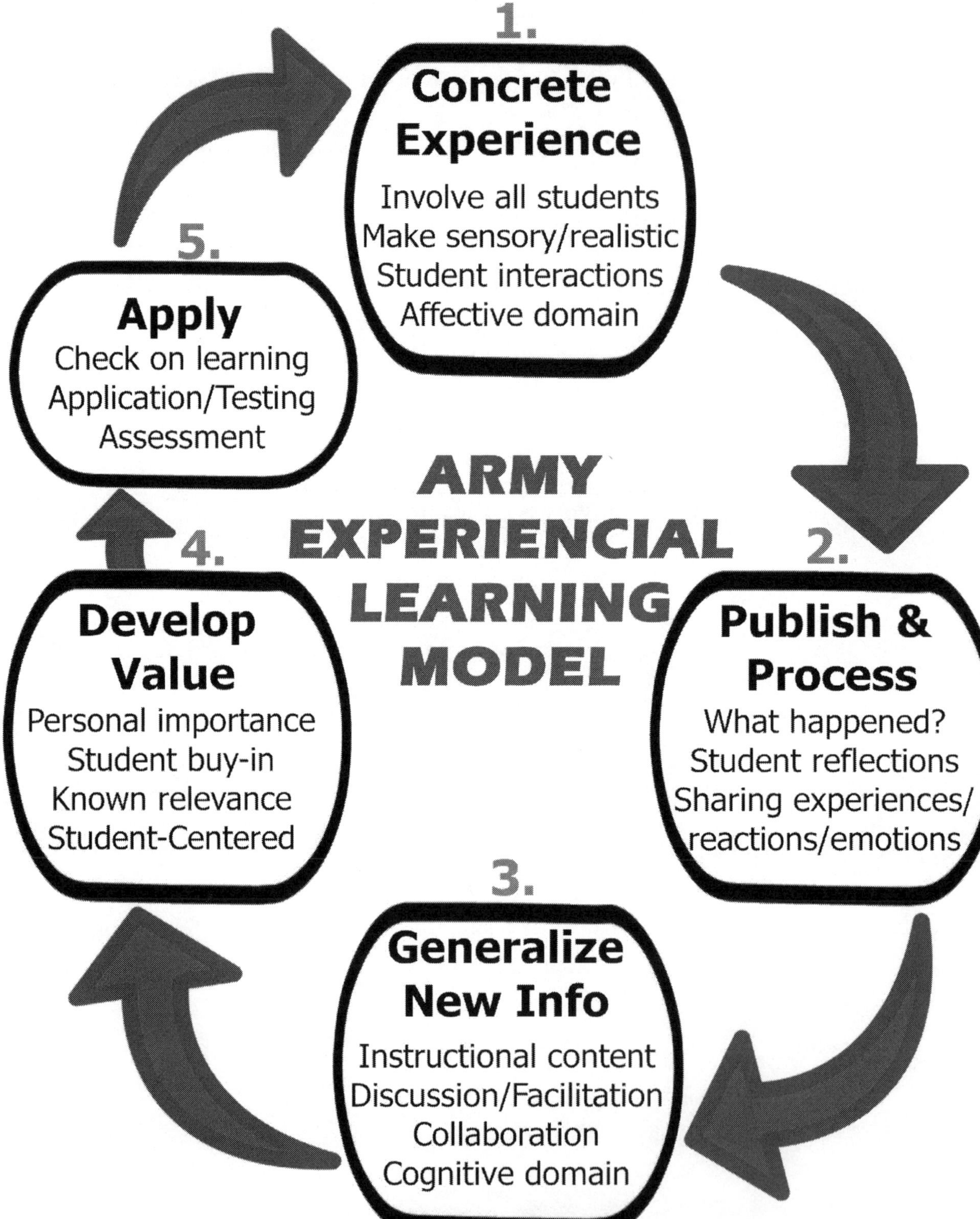
ARMY EXPERIENCIAL LEARNING MODEL
1.
Concrete Experience
Involve all students
Make sensory/realistic
Student interactions
Affective domain
2.
Publish & Process
What happened?
Student reflections
Sharing experiences/
reactions/emotions
3.
Generalize New Info
Instructional content
Discussion/Facilitation
Collaboration
Cognitive domain
4.
Develop Value
Personal importance
Student buy-in
Known relevance
Student-Centered
5.
Apply
Check on learning
Application/Testing
Assessment

Army Experiential Learning Model (derived from TRADOC Pamphlet 350-70-7, 2018)

In the Army Experiential Learning Model, a student first experiences a *Concrete Experience* event (and/or a trigger of prior experience and knowledge). Then the student goes through a *Publishing and Processing* phase where they react to the event through observations and reflections which are enhanced when shared with others. The student now goes through the *Generalizing New Information* phase. Dealing with the cognitive domain, this phase focuses on the content and methodology where the instructor provides additional instruction to help the student understand. The *Develop* phase deals with students thinking and discussing how the information presented is valuable (relevant) and could be applied in different ways in the future. The final step of *Apply* focuses on actually implementing what was learned such as through a practical exercise or some other practical application (this would serve as a check on learning of the learning objectives), (Pierson, 2017; TRADOC Pamphlet 350-70-7, 2018).

The informal Army After Actions Review (AAR) process (generally conducted after any training event/concrete experience) can be used as a great tool to help facilitate the Army Experiential Learning Model as well as improve student-to-instructor interactions – cognitive presence. In an AAR an observer asks specific questions of those involved in the training/situation event. Compare the steps/questions of the AAR to that of the Army Experiential Learning Model:

AAR Step 1: What was supposed to happen – *A Concrete Event*

AAR Step 2: What actually happened – *Publish & Process (student reflections)*

AAR Step 3: What are some things that went wrong (negative aspects) and why – *Generalize Info (discussion/facilitation)*

AAR Step 4: What are some things that went right (positive aspects) and why – *Develop Value (student buy-in/relevance)*

AAR Step 5: Use what was learned to do it again – *Apply (assessment)*

Two other noticeable similarities between the Army Experiential Learning Model and the AAR are that feedback should not be delayed, and maximum student participation should be sought to ensure all are engaged, learning, and understanding the learning event (Training Circular 25-20, 1993).

Teaching Presence

Teaching Presence, which deals with the physical process of instruction, is comprised of three main aspects: *design and organization, facilitating discourse*, and *direct instruction*. All three of these aspects work together for whoever is conducting the teaching, whether that is the main instructor/facilitator or a student who is instructing that specific class (Garrison, Anderson, & Archer, 1999). Notice how these teaching presence aspects overlap/tie in with many of the other items already expressed within this book; effective instruction (via the ALM) is a matter of addressing all aspects of learning.

Design and Organization: this aspect deals with ensuring a properly designed curriculum with the integration of the best methods of instruction for the subject and audience (focus on active hand-on learning). This section also entails fully communicating course goals and topics of instruction to lower anxiety and ensure students understand what is happening. Similarly, providing clear instructions must occur such as using understandable rubrics and easy to follow syllabi. Teaching presence is also bolstered by providing due dates with periodic reminders and properly holding students accountable. Additionally, ensure that the student learning community understands how to properly interact on and offline (netiquette) to ensure a safe/positive community of enquiry.

Facilitated Discourse: deals with the instructor/facilitator maximizing discussion/interaction by identifying areas of agreement with a conversion to help ensure student buy-in of the topic in order to fully grasp relevancy. Similarly, instructors/facilitators must seek to reach consensus to help maintain a mutual understanding within the community of inquiry; this isn't necessarily everyone agreeing, but it involves being able to at least understand other's point of view. To have and maintain the community of inquiry all students should be encouraged to contribute and should be praised for their hard work/effort. The discourse environment/climate must always be worked on to ensure it is safe, fun, and ridicule-free to avoid students shutting down and becoming reluctant to participate. Efforts should be made to maximize motivation and discussion while always keeping students focused and on track. To ensure understanding active learning, low-stakes testing should be conducted often along with prompt feedback to address any misperceptions. This helps to give students confidence in their understanding and creates a positive learning experience.

Direct Instruction: this important aspect pertains to direct engagement with students which should always strive to be as engaging and motivating as possible. Instruction should use multiple examples along with stories to help students understand and build emotional connections to the information for long term retention. Additionally, interactions should always be cognizant of cultural considerations when instructing in a particular area or to a particular group. Finally, direct instruction must also take instructional technology like multimedia into account. Instructors/facilitators must be familiar with the technology to use it appropriately to increase realism and impact at the right levels to enhance instruction and student engagement.

COMPONENTS OF TEACHING PRESENCE

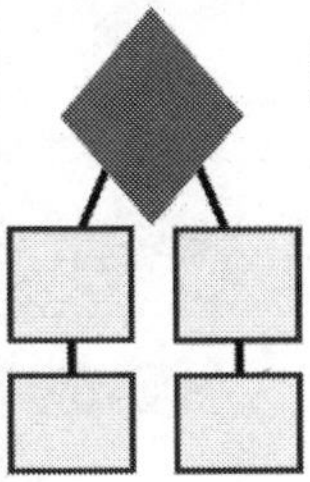

Design & Organization

- Properly design/communicate course goals/topics
- Provide clear instructions (*use rubrics & syllabus*)
- Provide due dates (*give reminders, accountability*)
- Explain conduct/participation (*proper netiquette*)

Facilitated Discourse

- Identify areas of agreement (*buy-in/relevancy*)
- Seek to reach consensus (*community of inquiry*)
- Encourage contributions (*praise hard work*)
- Set climate for learning: *safe, fun, ridicule-free*
- Draw in participants & keep students on track
- Focus discussion & seek maximum participation
- Confirm understanding (*low stakes testing*)
- Diagnose misperceptions (*timely feedback*)

Direct Instruction

- Present content/questions (*engaging/motivating*)
- Inject knowledge (*use examples and stories for an emotional connection*)
- Ensure that cultural considerations are identified & taken into account
- Use appropriate multimedia to enhance realism and fully engage the student

Derived from (Anders, 2017; Anderson, Rourke, Garrison & Archer, 2001; Shea, Li & Pickett, 2006)

Cultural Considerations

Cultural considerations (presented as an aspect of teaching presence) need to be addressed both when conducting instruction outside of the United States and when hosting international students. The Army has placed increased emphasis on this issue in that it has "…recognized the need for its forces to be culturally aware, culturally empathetic, regionally informed, and to use the appropriate language to facilitate communication and improve understanding with the nation's joint, inter-organizational, and multinational partners, and local populations" (TRADOC Pamphlet 525-8-2, p. 22).

Cultural issues such as differences in verbal/non-verbal forms of communications (differences in tonality and gestural meaning), religious practices (what can and can't be done on certain days/times), forms of dress, and information processing (greater reliance on context, implicit and explicit cues) need to be understood in order to ensure that instruction can be fully grasped and retained by the student (Samman, Moshell, Clark, Brathwaite, 2009). Additionally, Wlodkowski (2008), in describing characteristics of an excellent adult learner instructor listed expertise, empathy, enthusiasm, clarity, and *cultural responsiveness*, to highlight how important culture is in a successful educational endeavor.

It is strongly recommended to conduct cultural research by using resources such as embassies, CIA world factbook, the Center for Army Lessons Learned – CALL website, and Pre-Deployment Site Surveys (PDSS) when planning any type of international educational mission. Having traveled to a multiple of countries for deployments and military-to-military educational missions, I have experienced firsthand the difficulties that can occur when cultural differences are not fully understood and addressed, so I strongly recommend taking it all into consideration.

SIDEBAR: Cultural Considerations

I once had the opportunity to be sent to the country of Armenia as part of our National Guard State Partnership Program. They had requested special instruction dealing with how to better utilize their Noncommissioned Officers (NCOs), Sergeants, within their military. As a senior instructor for our state's Regional Training Institute, I was sent to conduct a week-long course on NCO duties and responsibilities. After talking to the Armenian military leadership and observing the students, I delineated that one of the main areas of attention that their NCOs needed at that time were additional leadership experiences. One way that I addressed this was through the process of preparing for and conducting a road march.

It was only a 6-mile road march, but it was a nice little challenge up in the beautiful mountains of Armenia. Their scenario was to conduct reconnaissance on an area partially through the road march. Given the number of student soldiers, I organized the group into a platoon with students playing the role of platoon sergeant and four squad leaders. This gave them the opportunity to conduct troop leading procedures and pre-combat checks/pre-combat inspections (PCCs/PCIs). Halfway through the road march, I realized that no one was drinking water even though it was close to 100 degrees and very sunny. I informed the student platoon sergeant the importance of keeping the soldiers hydrated and watched him properly spread the word to the student squad leaders who in turn, spread the word to the other students, but no one would actually drink water. My frustration grew as we walked in the heat and I contemplated why the soldiers weren't following orders to drink water.

I tried to analyze this situation from multiple points of view but it just didn't make sense to me. Just as I was about to yell at everyone for violating safety, I stopped as a memory popped into my head. I just so happened to be married to an Armenian. My beautiful Armenian wife and I have two little girls and we had often gone for walks in the park. The memory that I recollected at this time was of us walking in the park on a hot summer day. I watched as my daughters went to drink from their bottles of water, but my wife stopped them, "no, don't walk and drink at the same time. It will give you an upset stomach." When my wife said that, I thought "hu, I've never heard that before." I didn't say anything about it because I didn't think it was worth arguing about. I had pretty much forgotten all about it until that moment with the Armenian soldiers.

That was it; it was a cultural norm. Similar to how many "American" mothers tell their children not to go swimming for two hours after they eat, most Armenian mothers tell their children not to drink while they are walking. It wasn't so much about logic as it was about culture. This was just part of their belief and cultural norm. I told the student platoon sergeant to tell everyone to stop and take a break for just a few minutes. He did and as soon they stopped, everyone took out their water and started drinking. Imagine how the students would have felt if I had started yelling at everyone about drinking water while they were road marching, "this guy is crazy, you're not supposed to walk and drink at the same time," would have been their main take away from the educational event. Understanding the culture can make a huge impact on how one ends up interacting with and succeeding with students.

Student-to-instructor interaction is made up of many important components as described (verbal/non-verbal/written communication, instructional presence, cultural considerations). These items are all vital to the learning process in that they aid in motivation, comprehension, and the aspect of instructor approachability. If an instructor/facilitator can properly communicate in a motivating way that is open (approachable, not intimidating or condescending), students will be much more likely to ask questions and engage/interact with the instructor/facilitator which enhances the overall community of inquiry and therefore improves learning (Anders, 2017; Garrison, Anderson, & Archer, 1999).

STUDENT-TO-STDENT INTERACTION

Student-to-student or peer to peer interaction has great relevance to the real world due to how most missions/situations require people to interact and work together to address a dilemma. The Army highlights this importance, "Future Army forces require the capability to conduct peer-to-peer learning to facilitate problem solving, collaboration, information sharing, and provide virtual learning opportunities" (Training and Doctrine Pamphlet 525-8-2, 2017, p. 37). Additionally, learning research conducted by Jung, Choi, Lim, and Leem (2002) supports previous research that advocates "…for the importance of collaborative learning and social integration in enhancing learning outcomes, increasing learner satisfaction" (p. 159). Doyal (2011), in citing multiple other educational research, also stresses the importance of learning strategies that take the social aspect of learning into account and how those learning strategies tend to be much more successful because students are active participants (discussing, critiquing, offering their opinions/experiences) with ownership and collaboration in the learning. Wlodkowski (2008) also highlighted this position with regards to adult learners in stressing the importance of "engendering a feeling of connection among adults" (p. 134). To maximize student-to-student learning, issues such as how students are to interact with one another and how group work is to be conducted need to be fully comprehended.

How Students Should Interact with One Another

It is incorrect to assume that students naturally know how to properly interact with one another in a cooperative educational environment (as an experienced military instructor and college professor I've learned not to assume that all students understand proper class *etiquette/netiquette*). There are many reasons for students' unpreparedness in this regard such as: they are new or haven't been in an educational setting in a long time and have forgotten, the customs and courtesies in their workplace environment are very different than what is expected now, or they are from a different country so there are cultural consideration issues. Instructors/facilitators need to ensure that students can disagree politely (debating as opposed to arguing), properly ask for clarification (follow up questions), ask for reasons behind other's ideas, offer appropriate praise and thank others for their contributions (as well as offering *constructive criticism*) and be able to ask for assistance or suggestions from others as opposed to seeking to be given the direct full answer (Jacobs, Renandya, & Power, 2016).

Another important skill for many to learn when working in a group is the ability to not be overbearing/domineering; basically, to actually work as a team member instead of clashing heads, wanting to take control (Nilson, 2016). These skills and attributes should be modeled and emphasized by the instructor/facilitator from the beginning of the course with periodic reminders (as well as holding students accountable – instructors/facilitators must always have the proper balance of being authoritative while also being approachable - to ensure that group work is effective). Another idea would be to have a small test or assignment that directly covered this topic, and what better way for them to work on this assignment than as *group work*?

How to Conduct Group Work

There are many ways to conduct effective group work and active experimentation is encouraged. As an instructor/facilitator, using the Army Learning Model, one should always be trying new things in an effort to constantly improve the learning environment and process (Training and Doctrine Pamphlet 525-8-2, 2017, p. 24).

Nilson (2016) in stressing the importance of genuine learning, provides a poignant understanding of using group work within an educational environment:

> The group task must make students learn something, not just do something…and demand group synergy and higher-order thinking processes (application, analysis, synthesis, evaluation) to complete. … Group collaboration is the way the world works because well-functioning teams generate more innovative and creative ideas and devise better solutions to problems than do individuals with a competitive ethos. (p. 161, 165).

To help ensure that group work educational endeavors are impactful and successful, there are key areas to consider such as the size of the actual groups, along with learning strategy, the duties/responsibilities of each group member, and how to assess the group's performance/outcome.

Size of Groups and Learning Strategies

Group size is generally dependent on the type of activity being conducted. There are many different group work learning strategies, each with different amounts of optimal student participants. The following is a listing of a few effective group work strategies along with a short description of how to use them and the number of students to place within each group:

Think Pair Share: As the name implies this is generally done with a group of two students, but I have successfully done this with up to four students at a table. The instructor/facilitator presents a question or problem statement and first asks the students to simply think about it on their own (give students about a minute or two). Students would then discuss the problem with their pair/group, comparing answers (this usually takes 2 to 5 minutes – longer depending on the topic). Finally, each group shares what they have come up with dealing with a solution to the question/problem statement. It is often beneficial to have the group write out their final answer/solution (could be used for grading purposes or just to help the students obtain a consensus and organize their response). This is a great group work strategy for interjecting discussion into virtually any topic and is used quite a bit in higher academia.

Pros and Cons: Break students up into groups of four or six and assign half of them one side of an issue and the other side the opposing view. Give each side time to research, discuss, and organize either their pro or con view of the assigned scenario/issue. The overall group can then come back together to present their findings and see what the other side has come up with. It is again recommended to have each pro/con group side write out their final answers to assist in the assessment and with organization (be sure to tell students to come up with at least a minimum number of pros/cons depending on how they will be assessed). The times needed for each part of this group work depend on the subject matter and how in-depth you want them to go, be sure to also stress the importance of analysis, discussion, and critique as they work in their groups.

Scenario Creation: Divide students into groups of four or five and have them come up with scenarios that deal with a specific assigned issue. Have the students make the scenarios as descriptively realistic and personable as possible to maximize impact for the given situation (this could be done as a written or video creation project). The scenarios are to present a problem situation based on what has been learned from within the course, as well as their perceived solution. After groups have had time to make, discuss, critique, and refine their scenarios (generally the instructor/facilitator has reviewed the content as well) the students then give the problem scenario (without the solution at first) to another group which then comes up with their own response and then compares/critiques their solution to the original group's solution. This is an engaging and great group work learning strategy for stimulating creativity and higher-order thinking. Students also have a greater connection (ownership) to the learned material because it was created by them and other students.

Duties/Roles/Responsibilities Within the Group

There are many different ways to have students manage themselves within a group, but some key duties, roles, and responsibilities are generally needed. Depending on the size of the group, some students may have multiple duties, roles to play. Below are some different duties/roles examples for you as the instructor/facilitator to choose from for your students or to empower your students to select/utilize themselves:

Spokesman: group work should generally have a presentation component and assigning (or letting the group self-assign) a main spokesman is an important step. This isn't a leader position in that groups should be small enough where they can come to a consensus on decisions, this is just the main mouthpiece for the group (although having everyone take turns presenting is also recommended depending on assignment type and time allotted).

Writer: each group member should be keeping notes and writing their own portion of an assignment, but often an overall product will be needed from the group and it is important to identify who will be responsible for making that final written product.

Timekeeper/Organizer: this role is often tied together and deals with keeping track of time spent, time remaining, who has done what, group meeting times, and due dates. All should be keeping track of important times and dates, but it helps to have a person who has an explicit responsibility to manage the group's time.

Harmonizer: This position seeks to help the group in a multiple of ways to include keeping everyone focused and on track (a key learning process in group work is allowing for a lot of discussion, the harmonizer should work to keep the conversation moving in the right direction as much as possible). In dealing with conversation, another key role of the harmonizer is to help make sure that everyone's voice is heard. Some individuals may tend to want to dominate a conversation, so the harmonizer should help to ensure that all group members share and are part of the conversation. A simple exchange such as: "Thanks for your thoughts on that SGT Smith. We haven't heard from SPC Anderson about this yet. Please, SPC Anderson, tell us your ideas about this," can go a long way to ensure harmony within the group and have all group members participate and be part of the learning community of inquiry.

Liaison: this duty position is in charge of going to the instructor/facilitator with any group questions or concerns as well as to the other groups depending on what the assignment/project entails. Instructors/facilitators should always be going around observing each group, monitoring their progress, and clarifying any issues within the classroom.

Arbitrator: sometimes it is beneficial to have a designated arbitrator within a group to address issues of disagreement or to make a final decision if one can not be voted on. Again, this isn't necessarily a leadership role, it is just a position to help resolve situations so that the group can continue to move forward in a timely manner.

Group Assessment

There are many ways to conduct assessments for group work and in some cases group work might just be a quick learning process exercise as opposed to a fully graded assignment. A general recommendation in assessing group work is to give both an individual grade and a group grade assessment. In this way students will be directly evaluated and will have an understanding that they will not be able to simply get by in having the rest of the group doing all the work (meaning that they would be able to get a good grade if the group did well in spite of that student not doing their work for the group).

Another option is to have a portion of individuals' grades based on their own perceived assessment as well as peer assessments (peer assessments can also serve as a powerful tool in additionally motivating students to do their fair share of the group work). Assessment options and implementation depends on student level, type of subject matter, and type of course being conducted. Below is an example rubric for a group project:

Point Value	Excellent 5	Good 4 - 3	Developing 2	Beginning 1
Group Participation	Participated in all group activities (all classes/meetings). Assisted the group in achieving all of its objectives. Fully completed assigned duties, roles, and/or responsibilities given on time.	Participated in most group activities (attended at least 80% of classes & meetings). Mostly completed assigned duties, roles, and/or responsibilities given on time (at least 80%).	Participated in most group activities (attended at least 70% of classes & meetings). Mostly completed assigned duties, roles, and/or responsibilities given within group on time (at least 70%).	Participated in most group activities (attended at least 60% of classes & meetings). Mostly completed the assigned duties, roles, and/or responsibilities given within group on time (at least 60%).
Communication & Interaction	Always actively listened to others and used effective communication skills both face-to-face & online (proper etiquette & netiquette). Always encouraged group members to actively participate & share their ideas. Did not dominate group with own actions (shared power)	Usually actively listened to others and used effective communication skills both face-to-face and online (proper etiquette & netiquette), (at least 80%). Usually encouraged all group members to actively participate and share their ideas.	Actively listened to others and used effective communication skills both face-to-face and online (proper etiquette & netiquette), (at least 70%). Usually encouraged all group members to actively participate and share their ideas.	Actively listened to others and used effective communication skills both face-to-face and online (proper etiquette & netiquette), (at least 60%). Usually encouraged all group members to actively participate and share their ideas.
Work Contribution to Group	Excellent contribution, above & beyond required; work was excellent & required no revisions, additions	Good quality work; few revisions or additions are necessary	Poor quality work; substantive errors; much revision and editing is required	Poor quality; little, if any, contribution to group goals

Example of a Group Work Rubric *(modify and implement for your own use)*

Whenever conducting group work, it is important to fully explain to the students why group work is being done as opposed to work done individually (be sure to talk about the real world need to be able to effectively learn from one another and work as a team - *relevancy*). Ensure that students know how they will be graded/assessed within the group (individually and as a group). Additionally, provide specific instruction (both orally and in writing) on how they are to work together to complete the assignment (clear instructions are paramount, especially in a group work situation). Along with clear instructions, be sure to enforce timelines and due dates (often other group members or other groups are all waiting on and counting on others to complete their portions on time).

Why some people hate group work

There are people who very much dislike group work because they have had negative experiences in the past or simply prefer to do everything on their own. To address this issue stress how the assessment portion of the group assignment has been designed to prevent "freeloaders" and ensure that everyone does their fair share and is graded accordingly. If need be, you could even offer a group the ability to "fire" (kick out) a member who is not doing their portion of the group work (this is an extreme option but, depending on the type of class, could be a valuable technique). Also, stress how group work more closely matches how students will need to function in the real-world at their actual jobs/missions where they will most likely have to work in teams. With regards to the Army specifically "…combined arms teams must also be skilled in team building [group work] to maximize combat power" (TRADOC Pamphlet 525-8-2, 2017, p. 24). Even if an individual student is exceptionally smart and talented, they should never be doing all of the work on any type of project. Teamwork, helping and learning from one another, are important skills that students must all master to be successful in the long run; a lot more can be accomplished by a group working together than by an individual person.

STUDENT-TO-INSTRUCTIONAL CONTENT

This type of interaction refers to the student working directly with the instructional material. Student interaction with the instructional content could be through text, video, audio, or any other type of media (simulations, virtual reality, augmented reality, etc.). Key considerations with how students will interact with the content are how it ties in with assessments (assignments) and ensuring that the instructional content presented is relevant, challenging, and memorable.

Assessment – The assessment itself could be the central learning mechanism. In addition to hands-on learning (with hands-on assessment), any type of assignment can function as the actual learning mechanism. As an example, I once attended a high-level university course dealing with learning theory. The instructor gave a lecture and introduced a lot of big words and information. At the end of the lecture I had a bunch of new information, but I didn't understand anything. The instructor then gave us an *assignment* where we had to apply the theory presented in the lecture to a learning experience from our own past. To complete the assignment (which was a 7-page essay), I had to review my lecture notes, read and reread our textbook, find additional sources to cite, as well as reflect on my own experiences. At the conclusion of the assignment I had learned a lot about the learning theory, how to apply it, and I had incorporated this information into my own previous experience (previous knowledge). All of this in spite of the boring lecture which I had originally received. Notice, however, that the assignment required me to deeply interact with the instructional material, to understand, reflect, and then to apply it in a new personal way. As an example, recent research dealing with the use of assignments as the main learning mechanism for medical students have also shown to be very effective (Gajjar, Mirza, Shah, Mistry, 2017). With this understanding of the power of assignments in mind, assignments should be given that forces students to reflect, analyze, apply the learned information, and create new ways of presenting the knowledge that was gained.

Assessment should also closely match the way the instruction was given, especially with regards to low-stakes testing. In the previously provided example class dealing with how to properly emplace a tourniquet, low-stakes testing was done (all the students worked in a competition to be able to emplace a tourniquet in under 60 seconds). They were then tasked (assignment) to emplace the tourniquet in a harder, more realistic way (while wearing protective equipment and after moving the causality); the criteria/assessment requirement. However, was still the same: to emplace the tourniquet properly in under 60 seconds.

The purpose of creating assignments and assessments is to help achieve and measure educational goals and standards; with this regard it is helpful to organize and categorize the different learning objectives into an understandable and implementable standardized framework such as Bloom's Revised Taxonomy (an update on Benjamin S. Bloom's original catalog), (Krathwohl, & Anderson, 2009).

Cognitive Domain

The accompanying Bloom's Revised Taxonomy image presents the educational objectives dealing with the cognitive domain (mental capability/knowledge) from lower complexity (at the bottom: remember) to most complex (at the top: create). The six different levels within Bloom's revised taxonomy from simple to complex are 1. Remember (recognizing, identifying, recalling, retrieving), 2. Understand (interpreting, exemplifying, classifying, summarizing, inferring, comparing, explaining), 3. Apply (executing/implementing), 4. Analyze (differentiating, organizing, attributing), 5. Evaluate (checking/critiquing), and 6. Create (generating, planning, producing).

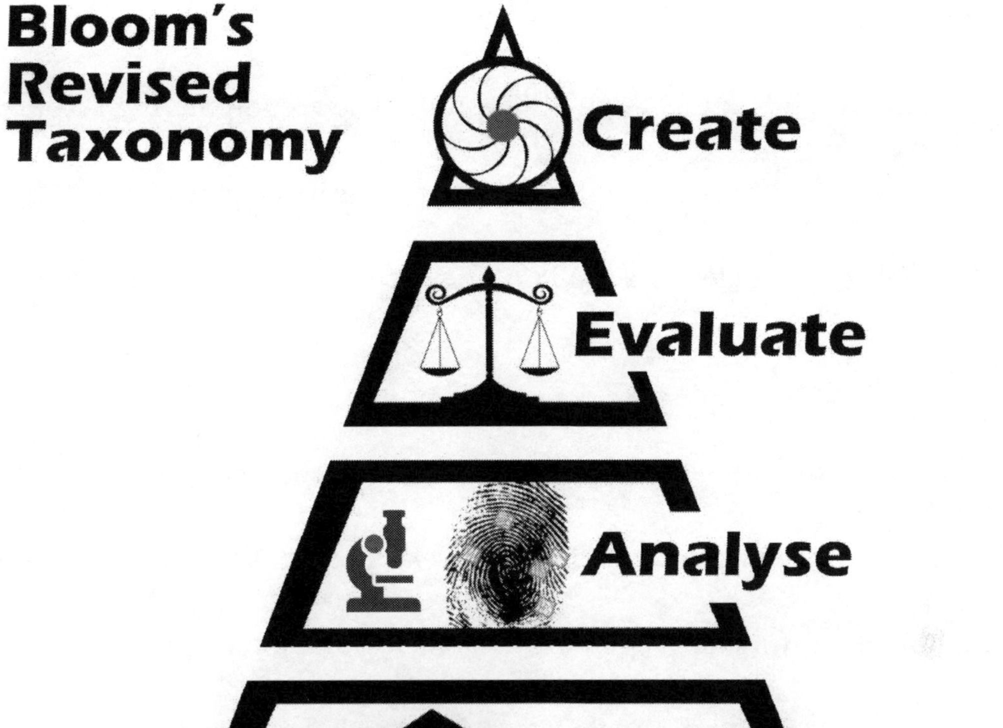

Bloom's Revised Taxonomy derived from research by Krathwohl and Anderson (2009)

Additionally, all of these educational objectives (cognitive processes) can be further categorized within different knowledge dimensions. The accompanying taxonomy table image displays the four different knowledge dimensions in a way to help instructors/facilitators more clearly see all of the possibilities with regards to class/course learning objectives.

The four different knowledge dimensions are: 1. factual knowledge (terminology, details, symbols), 2. conceptual knowledge (interrelationship information within a larger structure and how items function together), 3. procedural knowledge (ways/methods/techniques), and 4. metacognitive knowledge (process of thinking, awareness of one's self-own cognition - reflection), (Krathwohl, & Anderson, 2009).

The Knowledge Dimension	The Cognitive Dimension					
	Remember	Understand	Apply	Analyze	Evaluate	Create
Factual Knowledge						
Conceptual Knowledge						
Procedural Knowledge						
Meta-Cognitive Knowledge						

Bloom's Revised Taxonomy: Knowledge Dimensions (Taxonomy Table)

By using Bloom's Revised Taxonomy and taxonomy table, educational goals can be developed and become standardized when writing out learning, teaching, institutional, course, and class learning objectives (Krathwohl, & Anderson, 2009). These standardized learning objectives then become that much easier for purposes of organization and to ensure that all appropriate cognitive processes and knowledge dimensions are addressed so that there can be shared understanding between instructors/facilitators, students, and administrators (Krathwohl, 2002). Ensure that all learning objectives are written in plain, easy to understand language with action verbs that are properly observable, measurable, reliable, and verifiable to ensure attainment of the desired state. For additional guidance on writing military learning objectives see TRADOC Pamphlet 350-70-1, *Training Development in Support of the Operational Training Domain* (2019).

Although Bloom's Taxonomy is a powerful tool, many also use it to address all learning domains and stretch some of the actual meanings of the cognitive dimensions (example *Understand* to imply an emotional feeling or *Apply* to mean a physical process), Bloom's Taxonomy does not address the other domains of affective, and psychomotor, yet all are important elements of complete learning. "Educators need to effectively integrate learning strategies into all three learning domains to enhance the learning experience," (Johnson & Ferguson, 2018, p. 6). Even the researchers who created the revised taxonomy stated: "…nearly every cognitive objective has an affective component," (Krathwohl, & Anderson, 2009, p. 258). I will go even further and echo the belief of famed taxonomy of psychomotor domain developer Dr. Anita Harrow that whenever an individual is making a purposeful movement (psychomotor domain), they must also coordinate the areas of cognitive and psychomotor domains; one has to have knowledge (cognitive) of how and when to do something as well as see the value/need (affective) of doing something, and then have the ability to actually do (psychomotor) something (1972).

Affective Domain

This domain deals with feelings, emotions, and attitudes which relate to how we value, appreciate, and are motivated (Krathwohl, Bloom, & Masia, 1964). The affective domain is separated into a taxonomy of 5 different levels of internalization, from barely aware to part of one's feelings/emotions to full awareness and integration with behavior (Hoque, 2016; Krathwohl, Bloom, & Masia, 1964).

1. *Receiving*: this is having a sensitivity to or a general awareness of feelings, emotions, and the ability or wiliness to pay attention to it (lowest level of internalization).
2. *Responding*: deals with the active participation/attention of the learner and willingness to respond.

3. *Valuing*: refers to attitudes and beliefs regarding perceived worth – a commitment/preference for perceived value.
4. *Organizing*: the ordering/prioritizing of beliefs/value into an organized system; one thing is more important/valued than another thing: a value system.
5. *Characterization*: highest level of internalization of values/beliefs. At this level the student has now fully internalized to the point where learned values/beliefs affect their behaviors/attitudes, part of their internal philosophy of life.

The learning of all subjects has an affective domain component. We as instructors/facilitators must realize and remember that we aren't simply inputting data into a computer, we are working with humans who have their own set of beliefs and values who need to be convinced and motivated to understand and learn.

Psychomotor Domain

This domain deals specifically with kinesthetics, functional physical movements and reactions (with or without activities). Addresses the ability to manipulate one's self to accomplish a process, task, activity. Researcher Dr. R. H. Dave (1970) provides a taxonomy of the psychomotor domain divided into 5 discrete levels.

1. *Imitation:* observing and then being able to copy someone else's movements/behavior.
2. *Manipulation:* ability to follow guidance from an instructor/facilitator to perform an action/skill.
3. *Precision:* being able to properly perform (correct accuracy, proportion, and exactness) the needed action/skill from memory without assistance.
4. *Articulation:* able to perform two or more skills (sequenced and/or combined) in a consistent manner.
5. *Naturalization:* able to easily (automatic, with little physical/mental exertion) perform two or more skills (sequenced and/or combined) consistently.

It is important to note that this model/taxonomy of the psychomotor domain is specifically more applicable to higher education and has merit when applied to modern business and social skills in general which become part of virtually all subjects to be learned (Richlin, 2006). Though many view the psychomotor domain as only pertaining to very physical aspects, such as the movements required for Army combatives (hand-to-hand combat) or how to physically fix an engine, it has many additional implications such as body language congruency in communication and other social/cultural skills. Although the Army's official stance is that the psychomotor learning domain is to be regulated to the training environment, not educational institutions (TRADOC Pamphlet 350-70-7, 2018, p. 25), I believe ALM necessitates that all learning domains are taken into account within all training domains (institutional, operational, and self-development). As an example, look at one of the central aspects of the Army Learning Model, *critical thinking*. In order to fully learn about critical thinking one needs to understand lots of information dealing with problem-solving (cognitive domain), and being able to place value on different inputs of information and decisions or courses of action (affective), as well as being able to take action with that decision via persuading/communicating the plan to leadership and others (psychomotor).

All learning domains need to be considered and integrated when developing learning objectives to assist the student in obtaining full understanding. By having effective learning objectives, proper learning assignments/assessments can be created to ensure the most beneficial student-to-instructional content interaction.

Relevant: Although the aspect of relevancy was already discussed within the section dealing with motivation (relevancy was described as the "R" portion of Keller's ARCS Model of Motivation), it is worth readdressing with regards to how it directly ties with student-to-instructional content interaction. All students (especially Adult Learners) want the content they are learning to be relevant; no one wants to waste time learning something that is meaningless (Knowles, Holton, & Swanson, 2005). Another noted researcher Merriam, stated: “Assisting students in connecting learning experiences to their personal lives helps them to learn in ways that are relevant” (2001, p. 79).

Instructors/facilitators MUST work hard to ensure that students understand the importance/relevance of what they are learning. This means that it isn’t enough to simply be implicit when expressing the importance/relevancy of instructional content – instructors/facilitators need to be *explicit* and ensure that each student understands. This is more than simply making a “motivational statement” at the beginning of the class, it means really thinking through and finding impactful ways that the instructional content becomes personally relevant to each student (think personalized story, up to date information – recent news story, or creating an actual experience so that the student can really feel – emotionally experience the relevancy).

As a personal example, I remember being in Basic Training and learning about the maximum effective range of the M16A2 Army rifle (550 meters – point target). I heard a fellow student soldier say “Ugh, why do we have to memorize this for? We’re never going to have to actually use this information." Of course, he didn't say that loud enough for our instructors (Drill Sergeants) to hear because that would have resulted in doing multiple push-ups (punishment for talking out of turn and being negative). This is interesting though, in that this student wanted additional relevancy, whereas I was simply motivated to learn anything they told me so I could finish Basic Training (individual differences). I was highly internally motivated whereas this student soldier wanted more external motivation via more explicit relevancy.

A personal story by the Drill Sergeant explaining how knowing the maximum effective range of the M16A2 was vitally important in battle or possibly a virtual reality simulation in which a student could personally experience a situation where knowing the maximum effective range of the M16A2 would be relevant and lead to success (such as with tactical emplacement of troops for support by fire or for flanking maneuvers). This would all help the student see the relevancy, thereby be motivated to not only learn the information but retain for later use. Research with college students has shown that the use of simulations to help convey the relevancy of instructional content not only proved very successful but helped to *energize* students in being more interested in the content being learned (Montgomery, Brown, & Deery, 1997).

Challenging: This primarily deals with motivation and the *zone of proximal development*. This is directly relevant to students in that it also ties into the realities of what will be expected of them in their specific job/mission.

If the instructional content is too easy students will not feel challenged and can lose interest/motivation. In the same way, if the content is too challenging the student will experience cognitive overload (information is too excessive, complicated, and/or demanding) and will not be able to learn (Anders, 2017; Reynolds, Roberts, & Hauck, 2017). The difficulty is in finding the right zone of proximal development to help push the student to reach their full potential (Harland, 2003).

To achieve this right level of making the instructional content challenging, the content itself should be continually evaluated, assessment scores should be reviewed, and incoming students should be assessed to ensure that the content is at the right level and that students are ready for the instructional content. If an entire class of students are not ready, then the content should be modified or additional classes given. If a few students are not ready, then additional supportive *learning scaffolding* may need to be provided. Learning scaffolding is instructional techniques/content/assistive tools to temporarily help learners until they can either become more independent learners or have caught up with the rest of the class (like real physical scaffolding it is meant to be used temporarily and then removed when no longer needed).

Learning scaffolding can include additional instructional videos, text, presentations, quizzes, games, simulations, as well as tutoring and/or artificial intelligence assistive tools. An additional way to help reduce cognitive load (and avoid cognitive overload) is by chunking information. Chunking is organizing content into smaller more manageable and understandable bite-sized pieces. Examples would include explaining a concept one step at a time instead of the overall concept at once or presenting an online lecture video in logical 10 to 15-minute segments instead of one 50 minutes or two-hour-long video (Anders, 2015; Johnson 2017; Sousa, 2011).

As an example of scaffolding, if I were to teach a class on advanced map recon and tactical troop emplacement, I could have student-soldiers that are not familiar with or have forgotten how to read a map. I could create learning scaffolding for them by having a hand-out addressing simple map reading, or by creating an online course module with a presentation, video, quizzes, and games that would familiarize students on reading a map (available to students outside of and before class – and accessible as often as they need it).

The easiest solution isn't always best, and it can be challenging to find the full right answer; students must be able to fully think through and critically in any situation.

Memorable: To make instructional content more memorable it must be impactful and have an emotional connection. Content becomes impactful when it is relevant, understandable, and is presented in such a way that helps students become properly immersed in the learning process. Being properly immersed involves the use of multiple senses (multimodality) and creating an emotional connection to the instructional content.

The best way to provide instructional content depends on the subject matter and the type of students present (are they advanced or new). Little evidence exists for individual learning styles (we all have evolving preferences but even that is subjective); instead, the focus should be on presenting information via multisensory means (visual, auditory, touch, kinesthetic - movement, taste, olfactory-smelling), (Cuevas, 2015). In using the previously presented example class on applying a tourniquet, the hands-on means of instruction: being able to handle the Combat Application Tourniquet™ (CAT), then following along with the instructor, and finally doing it at combat speed, is a much better, more engaging, and impactful way to interact with instruction than simply hearing about it through a lecture. Given that this example is an action-oriented subject (students must be able to physically apply a tourniquet) it makes logical sense to use a kinesthetic means of instruction using movement, sight, auditory, and touch as important senses for learning. Additionally, the example class had a motivating guest speaker first, then students practiced on themselves, then on one another, and then all engaged in a competition to see who could do it the fastest. This type of interaction/engagement would result in real experiences with real emotions creating a greater sense of ownership and impact on the students resulting in better long-term learning.

IMPLEMENTATION EXAMPLE: Interaction

Yet what about learning content that isn't necessarily action oriented? Class topics that aren't physical/kinesthetic in nature can at first seem hard with regards to making it more engaging and hands-on. It takes effort, time, and creativity, but it can be done. The proceeding is an example of how I addressed this situation.

I once taught a class on critical thinking which was a very information-heavy type of class. Due to the somewhat abstractness of the class, it wasn't physically demanding like the previously described tourniquet class. The way I addressed this issue was by *making* the class more action oriented. After giving them a short initial presentation defining critical thinking and being able to move beyond linear thinking, we had a *practical exercise*. I broke up the class into groups of four or five and informed them that a new situation had occurred (this was the practical exercise's scenario).

Scenario: The Philippines, an Asian country consisting of many islands in the Pacific ocean, had just been hit by a hurricane and was requesting our (the Army's) assistance in re-establishing order and meeting the needs of the people (I left it generally vague but gave more detail to students when asked). Their assignment (action) was to work as a team to decide what they would do to address the issue.

I gave them a worksheet with previously learned operational environment variables: political (type of government/org), military (capabilities), economic (strength), social (culture/religion), information (data distribution/media), infrastructure (capabilities/resources), physical terrain (landscape/conditions), and time (weather/events/duration) (PMESII–PT) so that they could focus on the process of taking different things into consideration (and second/third order effects) as they used their critical thinking skills. I set it up so that the students could use their cell phones in one corner of the room to research information, they could also go to another part of the room to review the CIA World Factbook on that country (I had it printed out), I also informed them that I was acting as a subject matter expert on that region so they could come to me with questions and if I knew, I would tell them (they could also ask other students who had been to that country). I increased the emotional impact by telling them that "people had been killed, but many more people's lives were at stake," I also projected images of the devastation caused by the hurricane.

Image Credit: Lawrence Ruiz, CC BY-SA 4.0, commons.wikimedia.org/w/index.php?curid=43911464

There was lots of action going on in the classroom as groups generally assigned different members to research different variables. Students were up and moving around as they gathered information and put thought into what to do to address the situation presented. There was lots of discussion and debate going on as the groups finalized their creative plans and filled out a final version of their PMESII–PT worksheet. In the end, I had each group come up to the front of the class, present their courses of action, and answer questions about their plan to address the crisis.

So, what was the result? What did the students think of instruction presented in a way that maximized instruction by taking into account all the aspects described with regards to student-to-instructor, student-to-student, and student-to-instructional content? In reviewing end-of-course-critiques, students made statement such as: "I learned a lot and found the class really impactful," "I appreciate how it wasn't a boring lecture and instead we actually got to do something," "Usually I hate working in groups, but this was great," and "Thank you for making it emotional and fun!"

EMOTIONAL CONNECTION

That last student comment, previously mentioned, is particularly interesting in that the emotionally laden word "fun" was used. The word "fun" is a small word but has a lot of power, especially when dealing with education in that it indicates that a genuine emotional connection was made. Although the aspects of human emotion and learning are complicated and research is ongoing, there are some things that can be incorporated to enhance our educational environment. One key finding is that students having a positive mood has a positive effect on learning, but only if the content that the students are learning is viewed as being significant/relevant by the students (notice the overlapped importance with motivation as previously mentioned), (Abele, 1995; Hascher, 2010). Additionally, anxiety/negative emotions should be reduced and addressed, "…we should be more sensitive to processes of unsuccessful learning situations. Failure will stimulate negative emotions [anxiety] which will amplify the turning away from the learning situation. …students need positive emotions to stay on the task, to be open for feedback" (Hascher, 2010, p. 17). Students having some stress can be motivating, but too much leads to excessive anxiety which lowers motivation, prevents concentration, and prevents learning. Measured constructive feedback with both positive and needed areas to work on, as well as how to actually improve should be thoughtfully provided to the student by the instructor/facilitator.

Hascher, in citing work done by Valeski & Stipek, (2001) also addresses the importance and influence of instructor's/facilitator's emotional state and anxiety level, "teachers who enjoy teaching are more creative and more supportive during student learning... By caring about a positive social climate, teachers can foster the development of the academic self-concept of students and their motivation to invest in learning" (2010, p. 24). Additionally, research has indicated that instructor's/facilitator's enjoyment/enthusiasm for the teaching endeavor was to a large extent transmittable to the students, thereby enhancing their emotional state for learning as well (Frenzel, Goetz, Lüdtke, Pekrun, & Sutton, 2009).

Fun!

Going back to the specific notion of "fun," Lucardie (2014) in conducting research on adult learning and "fun" noted that having fun and enjoying the course "…were perceived by both learners and teachers as a motivator to attend classes…were considered a mechanism that encouraged concentration …helped in the absorption of learning…were identified as a proven way to build a socially connected learning environment" (p. 439). Stephen E. DiCarlo (2009), researcher, professor of Physiology and Claude Bernard Distinguished Lecturer, directly addressed this issue while giving a speech entitled: *Too much content, not enough thinking, and too little FUN!*

> We must create a joy, an excitement, and a love for learning. We must make learning fun. …If we are truly successful, we will have students who are engaged,

> inspired, and really learn. Students who are able to solve novel problems. … To achieve these goals, we must 1) reduce the amount of factual information that students are expected to memorize, 2) reduce our use of the passive lecture format, and 3) help students become active, independent learners and problem solvers. (DiCarlo, p. 261)

DiCarlo went on to express how most information being learned now will quickly be outdated and described how multiple research has shown that students forget a majority of the data they are forced to memorize for any given class (Cortright, Collins, Rodenbaugh, & DiCarlo, 2003; Miller, 1962; Swanson, Case, Luecht, & Dillon 1996). Students often memorize answers by cramming for a test and then *information dump* (forget everything) after a test (Doyle, 2011). DiCarlo's reasons for avoiding the passive lecture format were that again, multiple research has shown that students lose interest after 10 to 15 minutes and that passive lectures are simply boring and are not effective (Hammen & Kelland, 1994; Penner, 1984; Stuart & Rutherford, 1978). Finally, as to the reasons why DiCarlo stressed the importance of helping students become active, independent, problem solver learners, "…it is clear that active processing of information and not just passive reception of that information leads to learning. That is, we understand and remember the information we think about" (2009, p. 262).

As to how these goals can be achieved, DiCarlo expressed the importance of using active-learning strategies because they are engaging, *fun*, and "…reach all types of learners in the visual, auditory, read/write, kinesthetic, and tactile schemes" (2009, p. 262). DiCarlo also cited multiple research (Matlin, 2003; McDermott, 1993; Rao & DiCarlo, 2001) for his claim that "…students have better retention and understanding of knowledge when taught by active as opposed to passive methods" (2009, p. 263). For even more credence, Hascher (2009), in citing Schürer-Necker (1984, 1994), makes the powerful statement that "information needs to be physiologically [physically] activating and emotionally touching to be processed and remembered in high quality" (p. 17).

As a personal example of the additional power of the word "fun", I recently went through the United States Army Sergeants Major Academy. I was a bit worried about attending the final residency portion of the course. Although education is my life, I was still intimidated to attend the highest academic institution within the enlisted ranks of the Army. But about a week before I arrived at the Sergeants Major Academy, I got an email from the instructor that read "…we are excited to have you in our class. We will have fun throughout…". The fact that the instructor/facilitator used the word "fun" made all the difference in the world to me. It instantly lowered my anxiety and gave me confidence that they weren't interested in just trying to fill me up with facts to memorize but were instead wanting to provide us with great, usable educational experiences, all while having *fun* as well. The Sergeants Major Academy residency portion of the course ended up being a wonderful hands-on experience with great engagement, fun and was an extremely valuable educational event. They had recently revamped most of the way they conduct face-to-face instruction to be more in line with the Army Learning Model. All the instructors and administrators I talked to at the Sergeants Major Academy were true educational professionals (especially SGM Herring - thank you for a transformative experience).

For students to gain deeper learning by having enhanced fun engagements, we, as instructors/facilitators, must not only work to incorporate emotionally impactful teaching into our classes but must also have the right mindset. While attending a workshop dealing with improving instruction, the presenter strangely made the statement: "as a professor I'm not in the classroom to be a *babysitter* or to *perform*. I'm only there to *impart knowledge*." To make it even more strange, several professors in the audience shouted out in agreement with his statement. These professors had the wrong mindset in that they were mainly researchers who also *had to* conduct classes.

To be an effective instructor/facilitator one needs an enthusiasm for teaching and one does need to be an effective performer with excellent verbal and non-verbal skills (instructional presence). Although we as instructors/facilitators instruct adults and aren't *babysitters*, a good babysitter is also engaging and fun. Although we as instructors/facilitators are to be subject matter experts in our field, *imparting knowledge* is only part of what we do – we need to help the student understand how to properly think (critically/creatively), research, and address any given issue. We need to work to help transform the student into a better learner. We need to go through this learning process together (instructors/facilitators and students) in a community of inquiry. Being cognizant of making instruction "fun" is a powerful enhancing component to instruction; additional specific information on enhancing emotional connections and increasing fun is provided within the Instructional Methodologies section.

MOVEMENT & EXERCISE ENHANCES LEARNING

Exercise improves learning, it's that simple. This is an area that is often overlooked in academia but there is a lot of research to support this enhancement to learning; students who exercise improve their learning capabilities (Doyle, 2011; Jensen, 2000; Taubert, Villringer, & Lehmann, 2015). Exercise, in general, provides an overall effect of improved blood flow to the brain which helps with learning in general (Taubert, Villringer, & Lehmann, 2015). Recent specific research has shown that there is also great benefit in having students do active exercise for 30 minutes before class (Perini1, Bortoletto, Capogrosso1, Fertonani, & Miniussi, 2016), as well as exercise four hours after a learning session (van Dongen, Kersten, Wagner, Morris, & Fernández, 2016). Yet there is also plenty of research addressing the need to avoid instruction where all students do is sit and do nothing but listen; they need to stretch and move (Jensen, 2000). Additionally, a recent research study from the University of Stanford was able to show that "walking boosts creative ideation in real time and shortly after… Walking opens up the free flow of ideas, and it is a simple and robust solution to the goals of increasing creativity and increasing physical activity" (Oppezzo & Schwartz, 2014, p. 1142). This research gives yet another reason to avoid long passive lectures where students are trapped in their seats, forced to listen to a boring lecture. Instead, active experiential hands-on learning should be used as much as possible.

STUDENT CHEATING: Wanting to Game the System

This is an interesting section that I think should be thought about by all instructors and facilitators. Students will often want to game the system (finding easy solutions through help or feedback instead of learning material) and in many cases cheat in order to successfully pass a course. Research has shown that over half (some statistics show numbers closer to 75%) of students in higher academia have admitted to academic cheating at least once in one way or another (McCabe, Betterfield, & Tevino, 2012; Whitley, 1998). Research has shown virtually no difference in rates of cheating between online and face-to-face instruction (Watson & Sottile, 2010). I want to come right out and publicly state that I am very much against cheating to include plagiarism, but I also very much believe that in many if not most cases the student isn't the only one at fault.

In reviewing a multiple of research and in incorporating my own experiences and discussions with other military instructors and university professors, the top reasons students cheat is because of:

1. **RELEVANCY (motivation):** students don't see the importance (relevancy) of really learning the material presented.
2. **TIME:** students felt that they didn't have enough time to properly learn the material.
3. **ACADEMICALLY READY:** students were not academically ready for the course to begin with, so they felt they had no other option.
4. **ENVIRONMENT:** students were in an educational environment that didn't fully value academic honesty and learning.
5. **DIDN'T KNOW:** students didn't know or feel like it was cheating.

(Anders, 2018; Yu, Glanzer, Johnson, Sriram, & Moore, 2018)

Relevancy/motivation is a major aspect of the Army Learning Model and for good reason, in that, it affects everything that a student does to include cheating. As instructors/facilitators, we must always work to explicitly inform students why what is being learned is important and relevant both now and in the future. We must work to make the learning process as personal and meaningful as possible so that students see the value in learning and remembering so that they don't view it as a waste of their time and effort. We must help them see the value in studying the material because it is to their own personal benefit instead of allowing them the motivational excuse to cheat.

Time is another reason that students end up cheating on tests/assignments. Being able to manage one's time to properly study for a test and complete assignments is a personal responsibility, but we as instructors/facilitators can still help set up students for success in this regard. A clear syllabus with a detailed schedule, periodic important date reminders, ease to understand rubrics, study guides, study sessions, test-taking tips, practice tests/quizzes, online skill mastery apps/games, and timely/relevant feedback are all ways that instructors/facilitators can help students manage their time better.

Additionally, we must be sure that what we are teaching, and assigning are important and will help the student and not be busy work or be so excessive as to be an unreasonable amount to accomplish in the time given. When everything becomes important and the number one priority then nothing is really important or a priority. Be realistic with requirements (are you giving them enough time to do the readings/assignment along with all their other requirements: other classes, missions, work, etc.) to avoid putting students in a situation where due to time availability they may decide that they have to cheat in order to succeed and accomplish the mission

(Anders, 2018; Wong & Gerras 2015). In this way, we give students reasons to succeed instead of excuses to cheat.

Being **Academically Ready** is an important prerequisite for succeeding in any course but at times students will not be ready and instructors/facilitators should be able to properly address the issue. Providing resources for outside academic tutoring is sometimes an option and should be expressed/provided when available. Tangible actions that all instructors/facilitators can provide is just-in-time learning and learning scaffolding. An example would be providing a separate online module with videos and other instructional content as well as practice quizzes to help bring the student up to speed, which is to make them academically ready to succeed in the class.

Environment meaning the climate and culture in which instruction and learning occurs, which will have a major impact on students' perceptions and possible propensities for cheating. This is an especially important aspect for me in that it is a direct reflection on the actions of instructors/facilitators. In interviewing a young student soldier for my research dealing with Army online learning (Army Structured Self Development replaced by Distributed Leaders Courses) he expressed his thoughts on whether leadership should be concerned about online cheating

> …morally and ethically, yes [leadership should be concerned], because you're cheating on a test. But at the same time, from a pragmatic point of view, no. Because again so few people actually retain any of that stuff. I, as a leader, would not be overly concerned because I know I'm just going to end up doing some congruent training with them later on when I have time. Or I'm gonna tell them "hey work on this and this while I'm not around." And I know they're gonna retain that [hands-on learning] a whole lot better and a whole lot more of this is going to be relevant to them that way… At the end of the day, the most important thing is the job got done. Anders, 2018

The mindset of "just get it done" is the problem that is often expressed with a lot of training in the Army and to a large extent within higher academia as well (Reed, 2018; Simkin & McLeod, 2010; Wong & Gerras, 2015). This is problematic in that promotion, advancement, or grades are viewed as the most important thing instead of knowledge and skills which are of real value and needed for success on missions and in the workplace. Leadership and instructors/facilitators must serve as role models and set the example of valuing the community of inquiry, the learning process, and educational accomplishments based on students' efforts, real learning, and gained skills/abilities.

"…one important reason why students refrain from cheating: the presence of a moral anchor in a faculty member whose opinion mattered" (Simkin & McLeod, 2010, p. 13).

Students expressed that they either **didn't know or didn't feel like they were cheating**. All students should be fully informed on academic dishonesty policies with full descriptions and links provided in course syllabi. Students from other countries/cultures might have different ideas on what is cheating (as well as plagiarism) so it is important that all are in accordance with course/school policies. An important consideration in this area is for instructors/facilitators to be congruent with what is being taught and how students are reacting. By this I mean if we teach the importance of teamwork and how in the "real world" we will have to work with others to accomplish a mission and complete our jobs, then are we being hypocritical by not letting students work together to accomplish a task/assignment/test? In the same way, if we teach students 21st-century skills like digital literacy, properly using reliable sources from the Internet, and critical thinking, yet then prohibit them from using a search engine from effectively and logically finding the answer to an assignment/quiz/test, are we not being hypocritical? It is hard to make a case against being able to "Google the answer" (use a search engine to find the answer) in an age where one can access the Internet from any location on Earth with one's laptop, tablet, smartphone, or watch.

Instead of testing students' ability to memorize the answer we should instead be testing students' ability to apply, analyze, evaluate, and ultimately create. "…change the assessment from objective measures (multiple choice and true-false) to more subjective (essays and research papers) that require a more in-depth understanding of a topic and more personal expression" (Watson & Sottile, 2010, p. 11). Online resources such as Turnitin, PaperRater, and PlagTracker are assistive tools to detect plagiarism/cheating. Another good way to help reduce plagiarism and cheating is by giving students different and more personal assignments as opposed to reusing old tests/assignments that could be copied and released on the internet for others to cheat from. All the exemplar aspects of the Army Learning Model (a focus on student-centered learning, ensuring student motivation-*especially relevancy*, using experiential learning, developing critical thinking) all work to help the student develop a greater personal connection and responsibility to the learning process which also helps to reduce cheating behavior.

OTHER INTERACTIONS

In addition to the three main types of interaction (student-to-instructor, student-to-student and student to content) there are other interactions to consider such as student-to-self and student-to-technology. As a facilitator/instructor, it is important to understand how these additional interaction types affect learning and how they can be enhanced through instruction and facilitation of the community of inquiry.

STUDENT-TO-SELF

Student-to-self is an often-overlooked interaction type even though research has shown that self-interaction (also known as reflection) is crucial to deep understanding and long-term learning (Nilson, 2016). Although reflection is a common word and has already been used in this document often, it is still important to have an operational definition of it and what is meant when referring to *reflection* (student to self-interaction /student self-reflection). Reflection is a process where one stops to contemplate and think deeply about one's own thoughts/memories/experiences, as well as information presented.

The importance of reflection was also expressed with regards to the Army Experiential Learning Model step 2, Publish & Process (which includes reflection), "education connects experiences and prior knowledge through reflective judgment to construct a new understanding of complex situations allows adaptive instructors to facilitate learning through the Army [Experiential Learning Model]" (TRADOC Pamphlet 350-70-7, p. 15). The Army also addresses the importance of reflection with regards to critical thinking "The key to developing effective critical thinkers is to develop leaders who are purposeful and reflective" (Training and Doctrine Pamphlet 525-8-2, 2017, p. 21). With regards to critical thinking, needed components of critical *reflection* when trying to address an issue/problem are: clarity (fully grasping the meaning), accuracy (avoiding distortions), precision (necessary level of detail), relevancy (pertains to this situation), depth (interrelationships), breadth (multiple viewpoints), logic (no contradictions), significance (importance), and fairness (justifiable, no bias), (Elder & Paul, 2016). The act of reflection needs to be fully recognized and integrated into all educational endeavors.

Student reflections can be enhanced by purposely addressing it within a course by including reflective-type questions within quizzes, tests, and written assignments as well as with special projects such as presentations or videos. A simple but useful example to help with reflection is called the "one-minute paper" (Wolff, Wagner, Poznanski, Schiller, & Santen, 2014). Here the student is given a minute (or two) to reflect upon and write their reactions to learning material/experience. This process allows the student to reflect upon the learning event, incorporate prior knowledge, and more deeply contemplate the information. Typically, one-minute papers aren't graded (participation grade could be given) but they should be collected and reviewed to see how students are thinking about the class. These papers can be used as discussion points allowing students the opportunity for self-assessment in a low-stakes environment. Variations of this example include the "muddiest point,' and the "golden nugget," where the instructor asks students to take a minute to reflect on the muddiest point (something unclear) or to reflect on what was the "golden nugget" (best thing learned) in that particular class period.

A similar technique I've employed, even with adult learners, is to write out their reflection using the minute paper, and then also have them draw an *emoji* (☺) to further illustrate how they feel about what was learned/experienced (Anders, 2018). This adds a layer of emotion and contributes a visual element to the learning. This can be taken further by asking students to share their thoughts and emojis and describe what they drew. Discussions with drawings in this manner can also enhance the instruction "fun" component while learning.

Another major way to invoke reflection is through discussion and actually physically talking about one's own experiences/opinions. In fact, fMRI (functional magnetic resonance imaging) scans by MIT (Massachusetts Institute of Technology) researchers have shown that a much larger increase in brain activity occurs when physically speaking (actively engaging in a discussion) as opposed to silently listening, as in an instructional lecture (Vouloumanos, Kiehl, Werker, & Liddle, 2001). By having all students participate in a discussion where they contribute their own thoughts and views, their brain activation is greatly increased because they have to reflect on what was actually asked, what they have learned, and what they believe. This creates an enhanced educational experience as they are truly part of a learning community which leads to them gaining ownership of the process, their own learning.

Student Self-Motivation

An important capability for all students to have/learn and for instructors to be able to impart deals with how to be self-motivated (intrinsic motivation). Self-motivation is the ability for one to be intrinsically motivated in spite of outside influences (or lack thereof), in other words, being motivated even when the instructor/facilitator isn't motivating. Educational intrinsic motivation (self-motivation) is a vital skill to be developed within students so that they can maintain persistence and frustration tolerance as they pursue their educational endeavors to achieve overall learning success (Alex, Boris, Vladimir, & Suzanne, 2003; Reynolds, Roberts, & Hauck, 2017). Some key items that can be used to help develop this skill within students are to work to develop a student's growth mindset, help students see the value of establishing their own educational goals, explicitly informing the student of the relevancy of the class and helping them learn to find their own relevancy in any instruction, and teaching them the importance/power of a positive attitude, resiliency, and neurogenesis.

Work to develop a student's growth mindset: A growth mindset means that the student attributes their intelligence to their hard work and effort in studying and learning a topic (Dweck, 1986). Always express this narrative when commenting on students' work: *great effort on this, I can tell you studied intensely and worked hard on this, it shows.* Another growth mindset narrative to use when interacting with students would be *that feeling your experiencing that this is hard to learn, is the feeling of your brain growing, that's a good thing!* Dweck's multiple and ongoing research has allowed her to confidently state that "…students who believed their intelligence could be developed (a growth mindset) outperformed those who believed their intelligence was fixed (a fixed mindset)" (2015, p. 1). Teaching students to have a growth mindset can be enough in itself for learners to be self-motivated to educationally succeed (Young, 2019).

Help the student see the value of establishing educational goals: Helping students properly create their own academic goals can have a powerful effect on students' educational outcome by serving as an additional motivational device (Covington, 2000). Instructors/facilitators should encourage proper goals setting for all students through the use of S.M.A.R.T. goals (Doran, 1981). S.M.A.R.T. goals stand for:

Specific: avoid being vague, use as many details as appropriate
Measurable: how will you know you are on the right path/have succeeded
Actionable/Achievable: Can this be done, what do you need to do to achieve this
Relevant: why are you doing this, how is it important to you personally
Timely (Time-based/Timebound): set a specific time for accomplishment

By helping students see the benefits of physically writing out and following their S.M.A.R.T. goals they will greatly increase their chances of staying motivated and achieving educational success. Use a simple S.M.A.R.T. goals worksheet, such as the following, to organize a student's thoughts into a good, motivating S.M.A.R.T. goal.

S.M.A.R.T. Goals Worksheet

Initial Goal Draft:

S Specific	
M Measurable	
A Actionable/ Achievable	
R Relevant	
T Timely	

Final S.M.A.R.T Goal:

S.M.A.R.T Goals Worksheet

Inform students of class relevancy, learn to find own relevancy: The importance of relevancy has already been fully described in that it answers the powerful student question of "why is this important?" In addition to explicitly informing the student to the study material's relevancy (both current and future importance), instructors/facilitators should help students to manage their thinking so that they can find their own relevancy regardless of the class they are taking or the actions of the instructor/facilitator.

Help students understand that they should ask the following questions and if it isn't answered by the instructor/facilitator they should continue to seek out the answer through their own research, reflection, talking with peers, superiors, mentors, or other subject matter experts. Help students understand that answering the question of relevancy is very important to them and should always be addressed in order to maximize their own motivation and connection with the material for long term retention.

- Why is this information important?
- How will I personally use this information in the near future?
- How will I personally use this information in the far future?
- Is this similar to something I have already learned? How is it different?
- What happens if I don't learn this information, how will that affect me?
- Does this new information apply to everyday life? If so, how?

Teach importance/power of positive attitude, resiliency, & neurogenesis: A positive attitude in learning is important because it helps students overcome negative/frustrating situations and helps a student actively persist in working to succeed in their learning: "Attitudes translate into actions, and students with motivation to learn… and are more likely to retain information beyond a class session" (Reynolds, Roberts, & Hauk, 2017, p. 26). An effective way for students to have and maintain a positive attitude for learning is through the use of *resiliency* (Training and Doctrine Pamphlet 525-8-3, 2011).

Resiliency training is used throughout the Army and deals with helping Soldiers become better able to bounce back from stressful situations and better overcome psychological adversity. Resiliency training mainly consists of helping Soldiers become more self-aware of their thoughts/feelings, self-regulating with their impulses/emotions, being more optimistic (noticing positive things in life, self, and others), improving mental agility (flexible/perspective thinking), identifying character strengths (or areas to improve), and connection (establishing positive relationships), (Reivich, Seligman, & McBride, 2011). Notice how these enhanced elements of thought can be used to help students in becoming more self-motivated critical thinkers.

In addition to many great resiliency techniques such as Hunt the Good Stuff (identifying positive things in your life) and Icebergs (identifying small negative issues that are actually part of larger problems), one powerful resiliency technique deals with overcoming the process that many students go through dealing with catastrophizing, called Put it in Perspective (Army Master Resiliency Training, 2014). This happens when students feel overwhelmed and start to only focus on the bad things and ultimate negative possibilities. This type of thinking leads to excessive fear and students giving up on learning. To avoid catastrophizing students need to recognize what they are feeling through reflection, contemplate the best possible outcome and how to get there (planning and preparation), as well as the worst possible outcome and how to deal with it as well as how to best avoid it (studying and asking for assistance), (Army Master Resiliency Training, 2014). All students can benefit from resiliency training to persist and succeed even when faced with stressful learning environments or negative thoughts/beliefs.

Many older adult learners have the false, negative belief that due to their age they can no longer learn. This is a false belief in that neurological research has proven that people well advanced in age can still learn due to neurogenesis (Kempermann, Wiskott, & Gage, 2004; Shors, Anderson, Curlik, & Nokia, 2012). Neurogenesis is a biological process where the brain (even an adult brain) creates new fully functional neurons which allow all students, even those in late adulthood, to continue to learn throughout life (Gage, 2002). There truly are no valid excuses; we should all be life-long learners.

Instill the idea of learning something as if students had to teach it: learning research has shown that when students are learning content with the expressed plan of teaching it to other students, they learn the content at a better, much deeper level (Bryson, 2016; Duran, 2017). Informing students about this phenomenon and encouraging them to always approach learning as if they were going to turn around and have to teach it to others will help students be more self-motivated to learn the instructional content. This is also another reason why creating individual or group projects where students will then have to turn around and teach the rest of the class can be very impactful in students' learning journey.

STUDENT-TO-TECHNOLOGY

Interactions that occurs between the student and technology are another aspect of learning that instructors/facilitators must address to ensure learning is optimized. How much technology should be used, is it effective, and how familiar are the students with the technology, are important questions to ask and understand. By understanding how students are using the technological aspect of a course (both in face-to-face and online courses), the instructor/facilitator can take steps to ensure an optimal educational experience to maximize student learning (Chew, & Kwok, 2014).

Ongoing evaluation should be conducted on the types of technology being used and how it is being implemented. How is this technology affecting the learning process? Can this technology be used in a better way? What are the student's reactions to using this technology? Ensuring that students know how to use the technology will greatly enhance the learning experience (Rothman, Romeo, Brennan, & Mitchell, 2011). All instructors/facilitators should have mastery of the educational technology to ensure a smooth, problem-free learning environment. Instructor/facilitator training is vital for effective instruction and student learning. Additional technology variables such as ease of use by instructor, by student, cost of purchase, cost of integration, cost of maintenance, should all be taken into consideration (within an eye towards Return on Investment: ROI) when choosing new technology (Barnieu, Morath, Bryson, Hyland, Tucker, & Burnett, 2016).

Some students may not need any additional technological instruction, but others might need a lot of assistance. Options to address this issue include offering additional assistance after class, using student experts to help other students, providing a handout to explain how to use the technology, and/or creating an online video that students can view whenever they need. A combination of all these possible solutions generally best addresses the issue. Note that there are likely to be instances where students are embarrassed or uncomfortable asking for technical assistance; it is best to place created handouts and videos online so students can additionally access them outside of class.

The purpose of instructional technology is not to simply be on the cutting edge of science or to simply look cool, its main purpose must always be to help address the desired learning outcome(s) of the course. Although the use of new "flashy" technology can pique a student's interest (motivation), this newness doesn't last and is short-lived. Instructors/facilitators should choose to use technology because it aids in communication (discussion/feedback), engagement, and helps to visualize instructional information. If instructional technology is being used that hampers learning or makes things harder, then it needs to be re-evaluated in why/how it is being used. A perfect example of this would be the use of slideshow software such as the popular Microsoft PowerPoint® (instructional technology).

SLIDESHOWS (PowerPoint)

Slideshow software such as Microsoft PowerPoint has gotten a bad name/reputation recently, not because of the program itself, but because of how it is often incorrectly used. The term *death by slideshow, death by PowerPoint,* connotes misery experienced by having to suffer through a very boring presentation. But in reality, it wasn't the slideshow/PowerPoint technology that was boring, it was the content and how it was used. Sadly, many instructors, professors, trainers, and facilitators incorrectly use slideshows. Instead of using it as an aid to help explain their instruction, it is often used as a crutch and serves mainly as a script for them to follow. Yet properly using slideshows to help explain information more graphically has been repeatedly shown to improve understanding, engagement, learning, and provide a better educational experience (Apperson, Laws, & Scepansky, 2006; Duffy & Henry, 2016; Durso, Pop, Burnett, & Stearman, 2011).

I recently taught a course on instructional technology for mid to high-level officers where one of the main areas instructed was the utilization of slideshows. They acknowledged that the *death by PowerPoint* phenomenon was very common in the Army and appreciated the information presented on how to combat this issue. This instruction was conducted through the use of discussion, presentation, and hands-on learning through the practical exercise of actually creating and presenting an improved slideshow.

Although the Army Learning Model generally calls for minimal use of slideshows, the Army still heavily uses slideshows in educational environments. It is important to stress that slideshow instructional technology is just a tool (a powerful/flexible tool) and like all tools, it can be used properly or improperly. It is also important to understand that in the Army we use slideshows in two different ways: to *instruct* and to *brief*. When used for *briefing* purposes, slide shows are usually used to show the Commanding Officer that all the subordinate units understand what is going on; generally, the Commander has already seen the information ahead of time, so the briefing simply serves as a final blessing. The information presented within this book deals with improving slideshows for instructional purposes/settings (although this information is generally applicable anytime slideshows are used). The following provides a summary of key research findings (as well as my own experiences) dealing with how to most effectively use slideshows to enhance instruction and learning. Keep in mind the importance of maintaining students' focus, how the content actually looks from the students' perspective, and the cognitive load being placed on the students.

Design: Examine the following master slide image of a very typical Army slideshow template and focus on the top and bottom portion of the slide:

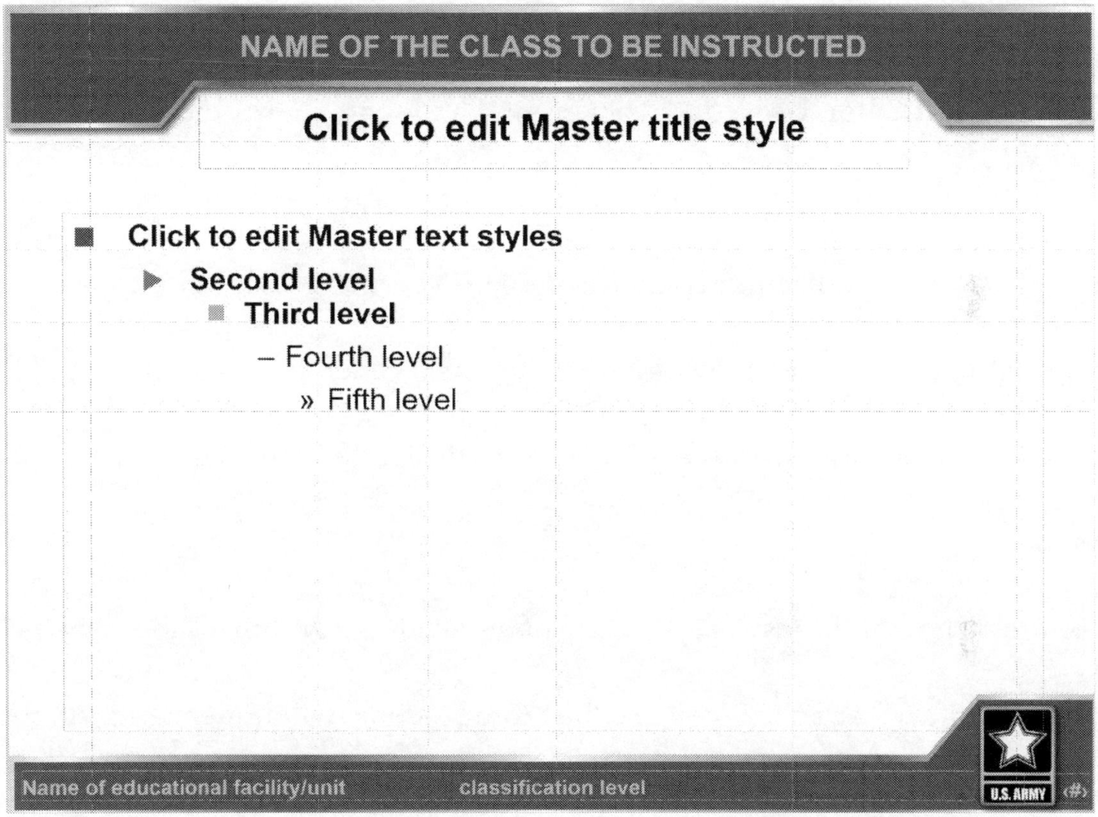

At least one-fourth of the slide real estate is taken up by these templated top and bottom sections. Although a consistent look and feel is nice, it can be accomplished through other means (via the same font faces and image/color styles). Information such as class name, educational facility/unit, and symbology could go on an introductory slide and then wouldn't have to be shown on every single slide. The same holds true with the classification level unless specifically needed due to regulations on sensitive information. This now allows for a greater amount of real estate (space on the slide) for actual instructional material. This then leads to the issue of the slide background color.

Background Color: Use a dark background and white/light-color text (the key is high contrast). This allows for a better focus on content, improved readability, increased contrast, reduced eyestrain, and avoidance of overpowering the viewer (Apperson, Laws, Scepansky, 2006; Temple, 2018). Additionally, this also generally makes most slides look more appealing (less boring), with a more modern design than the default white background.

Bullets: The simple answer to the use of bulleted text within a slide show is less is more, meaning the fewer bullets you use the more students will understand. Although a condensed bulleted text is way better than just putting up a bunch of text, it should still be kept to a minimum. There is competing research dealing with the maximum number of bullets to use per slide, but I now recommend using no more than 3 bullets whenever possible. Research across 10 different college professors with more than 300 students yielded the finding that "...the higher rated instructor used only an average of 3 bullets and 20 words per slide, whereas the lower rated instructors used 5-7 bullets and 25-70 words per slide" (Brock, & Joglekar, 2011, p. 89). This deals with reduced cognitive load (improved understanding) and helping to maintain student focus. Use logical judgment; if the information being presented deals with process or steps that would be easier to see on one screen then do so. Always think about how the information will be viewed and processed by the student.

When using bullets, instructors/facilitators should be sure to not present all information/bullets at the same time within a slide. Instead, each slide should be revealed one by one. In a large survey across a university and a separate private college, a large majority of the student expressed that: "The best things faculty members do with PowerPoint are to build each slide line by line and use bullet points with key phrases on slides" (Apperson, Laws, Scepansky, 2006). This aids with cognitive load and helps to properly direct student focus (Berk, 2012). If all the text/bullets are displayed at the same time, students will read all of the information on the slide while the instructor/facilitator is talking about bullet point, this will cause students to miss some of the instruction.

To allow for one by one bullet reveals in Microsoft PowerPoint, select the text box with the bullets, go to animation, and select "Fade." Then go to Effect Options, and be sure that "By Paragraph" is selected. To go even further with maximizing student focus and reducing cognitive load you can set your bullets so that the text of the bullet you just finished talking about will dim once you reveal the next bullet point. To do this in Microsoft PowerPoint select the text box with the bullets, go to animation, click on "Animation Pane," within the Animation Pane panel right click on your bullet and select "Effect Option..." Within the new popped-up window, under the default Effect tab there will be an After animation: selection that will be set to "Don't Dim." Use the drop-down selector to select a new color. If you had a dark background and had almost pure white as your text color, you should now select a slightly darker text shade and click the OK button. Now when you launch the slide show in presentation mode you will see that your bullets are revealed one by one and they each slightly dim after you go on to the next bullet point.

Having understood how to improve the use of bullets, one needs to really ask if bullet points should even be used; is there a better graphical way of presenting the concept/idea/information? International keynote presenter, graphic artist, and slideshow creation expert Gar Reynolds put it best by saying, "People remember visuals better than bullet points. Always ask yourself how you can use a strong visual—including quantitative displays—to enhance your narrative" (2012, p. 185).

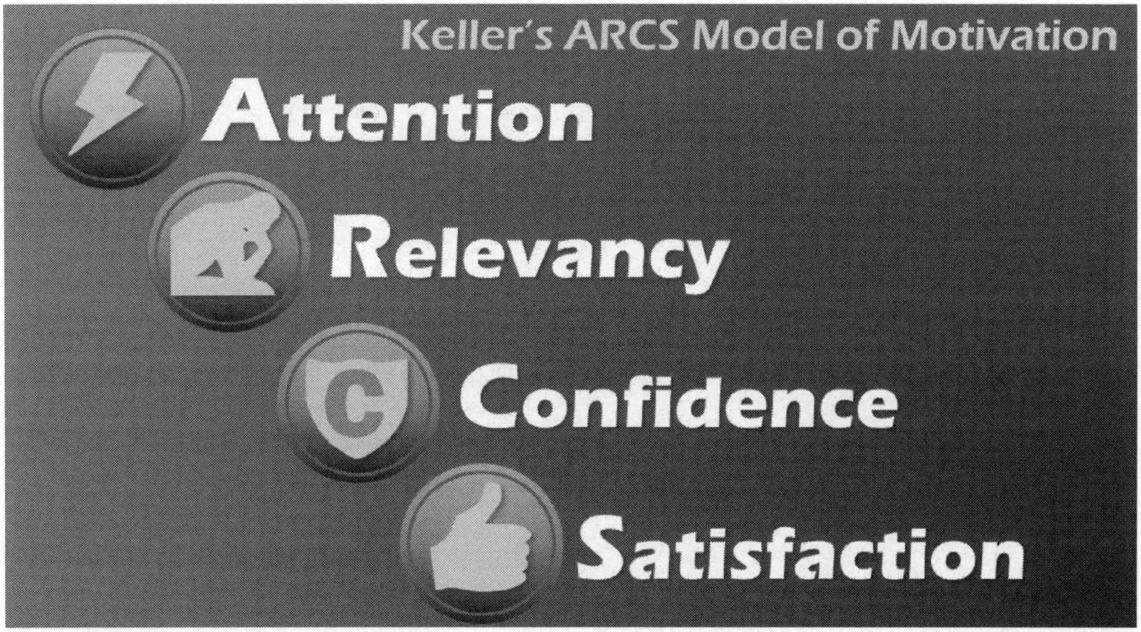

Graphics: Anyone that can see is mainly a visual learner so using enhancing relevant visuals will greatly improve any type of instruction (Merieb & Hoehn, 2007). Graphics are the singular most important aspect of a slide show, the graphic component. Think about it; the slide show's real purpose should be to serve as a visual aid for you as the instructor/facilitator in your instruction. Having said that, you really don't need a graphical aid of all the exact text that you just stated; in fact, multiple research has shown that students hate this (Kosslyn, Kievit, Russell, & Shephard, 2012; Selwyn, 2016). This has been specifically noted by college students in a large survey "…the worst things faculty members do with PowerPoint are to read each slide word for word without elaboration or discussion" (Apperson, Laws, Scepansky, 2006, p. 5). This also correlates with the cognitive theory of multimedia learning as described by Doyle (2011), expressing that greater education impact occurs when words are presented aurally in conjunction with associated images of the topic (not images of the words).

Ensure that graphics used serve a real purpose and are relevant to the instruction. Graphics should be high quality/clear, large, help to explain a concept, and/or evoke emotional tie-in with the instruction. Avoid using frivolous/busy images/animations in that if the image is not needed or is too busy/complicated it will distract the student and can lead to cognitive overload (Garrett, 2016). A simple but elegant and impactful image that properly correlates to what is being discussed can greatly aid students in understanding the material and retaining the concepts learned for later use (Roberts, 2018; Reynolds, 2012). An impactful image also sparks interest which enhances student motivation and can create an emotional connection. I strongly recommend using high-quality photos, silhouette images, and properly created graphs and charts as opposed to clip art and poor-quality images; this makes the presentation that much more professional and easier to understand. Compare the two examples of graphical choices below and judge for yourself which one is more impactful, memorable, and emotionally connecting.

Negative slide example of excessive text and clipart use, compare with the next photo example

Slide example of using an impactful photo to grab attention and motivation for course instruction

In the previous slide examples, the images represent the first slide of a slideshow dealing with establishing a Family Care Plan that soldiers must create when deploying if they have families that are dependent upon them. The first clipart slide is problematic in that the slide has a lot of text and the image used, clipart, does not really captivate or draw in the student. The second example uses minimal text and presents an impactful interesting photo. It is a current image of a soldier with their child (who looks like he might start to cry which creates even more emotion). The soldier pictured is perhaps getting ready to leave for a deployment, which ties in perfectly with the subject of the slideshow. This type of image would draw in the student and start to create an emotional connection with the subject matter due to the greater power and motivational effects of the photo used.

Text (beyond bullets): A vital element to making a great slideshow is to always take your students' view into account (student/learning-centered). By this I mean that you should literally set up the projector or digital screen with your slideshow in the room where students will be viewing it from, go to the back row of the class, sit down, and view your slideshow from there as if you were a student. Ask yourself how viewable will this be for students in the back row? Is the text size big enough? Are the colors displaying the way they looked on your computer screen? Is there enough contrast so that it is all easy to read and understand? Is there enough text or too much text?

There is still debate on how much text should be used, full sentences or just phrases (Brock & Joglekar, 2011; Garrett, 2016; Roberts, 2018). Most research and my own experience lead me to express that in general, less is more; use short phrases and then elaborate during your presentation. Consider the complexity of the topic being learned, your audience, and what other instruction (hands-on learning) will follow the slideshow instruction. Additionally, stick to one main message/concept per slide to aid in student comprehension, cognitive load (Seaman, 1998). Compare the following two images (slides) with regards to the text information presented:

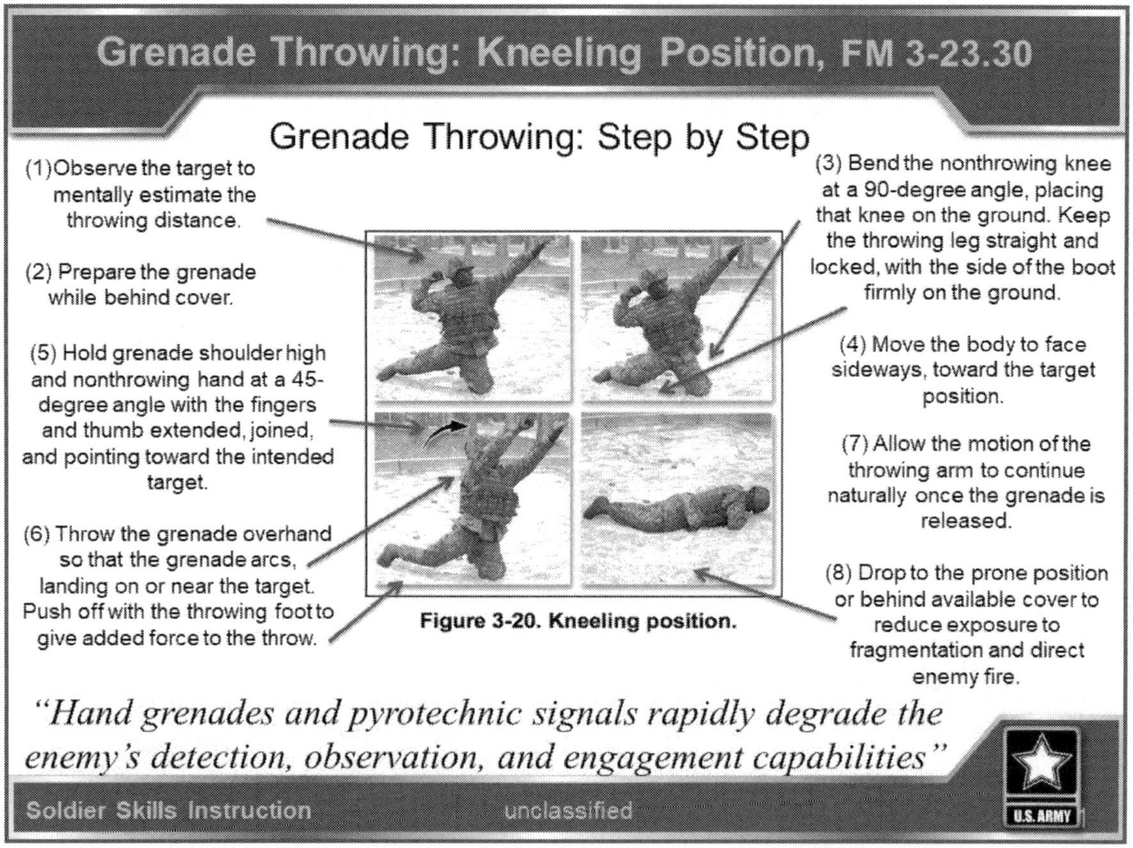

Slide example of what NOT to do (*excessive text/cognitive overload*)

Imagine presenting this slide to someone who has never thrown a hand grenade or hasn't thrown one for a long time. Most students would immediately start to read through all the steps (if they could see the small text) while the instructor/facilitator would still be describing step one. Additionally, if the purpose of this slide is to help show a student/soldier how to throw a hand grenade then why would the actual visual (photo image of a soldier throwing a hand grenade) be so small?

Notice how much space is taken up by an excessively large title area and bottom information section. Is the Army logo really needed there? What about the quote at the bottom of the slide? Does it really add anything to the instructional presentation? Sadly, this slide is indicative of the type of slides often presented within the military as well as within higher academia: too much text, small images, cognitive overload.

Compare the *bad example* slide with the following *good example* slide to see the difference in information highlighted. Remember to view it from the perspective of a new student who has never been exposed to information before.

Slide example of what you SHOULD do to avoid cognitive overload & maximize understanding

Notice how in this example a majority of the text has been removed. This information will be stated by the instructor/facilitator and he or she will be able to point out specifics as to foot placement and weight distribution by referring to the now enlarged image which is the central most important part of the slideshow. Title and reference information can also be removed on subsequent slides (subsequent steps in throwing a hand grenade) in that this information has now already been presented. Notice how all the focus is now on the image and what the instructor/facilitator is saying as opposed to the large amount of text that students would be reading which was presented in the previous example. Focus is a key element in helping students fully grasp the instruction. On subsequent slides, a red arrow or through the use of a laser pointer, specific focus could be brought to leg positioning to guide the student in understanding even more.

A defense I've often heard from instructors/facilitators is, "I have to put in a lot of text in order to explain the slides in case someone is viewing my slides later. The slideshow needs to stand on its own." I have two responses to this. One solution is to use the notes section of the main slide window when editing within Microsoft's PowerPoint. There is a text section labeled "Tap to add notes" (see red indicator arrow on image).

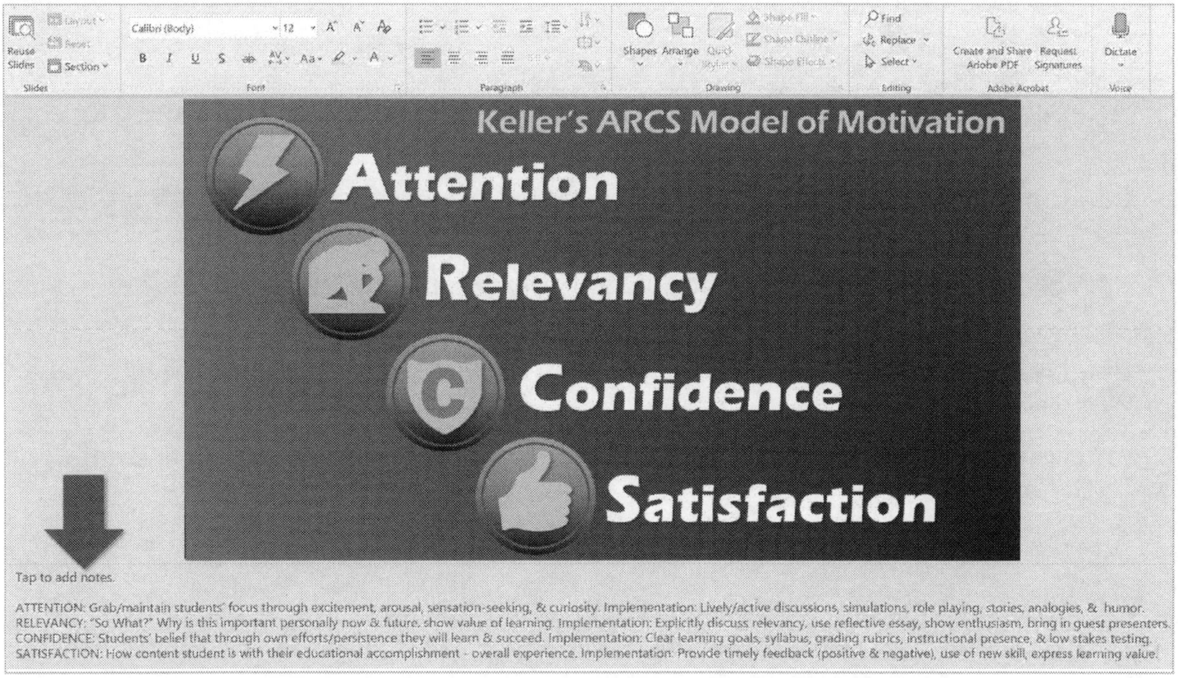

Using this notes section would function as a great area for additional explanatory text so that the main visuals of the slideshow do not get bogged down. A second solution is to create two versions of the slideshow, one with additional explanatory text, the other with very minimal text. The first one serves as a guide for you, the presenter, as well as a standalone slideshow. The second, more minimal text slideshow, would be what would actually be used in class (with additional explanation/elaboration given by you, the instructor/facilitator during the presentation).

Yet a slideshow itself doesn't serve as the best type of handout for the information being presented, there are better ways. Slideshow design expert Gar Reynolds expresses this important understanding,

> Attempting to have slides serve both as projected visuals and as stand-alone handouts makes for bad visuals and bad documentation. Yet this is a typical approach. PowerPoint and other digital presentation tools are effective for displaying visual information that helps tell your story, make your case, prove your point, and engage your audience. Presentation software tools are not good, however, for making written documents. That's what word processors are for. (2012, p. 71)

Give full consideration to the purpose of your handouts and the best way to distribute information if you are not physically presenting the information. There is also merit in encouraging students to properly take notes, either through providing an outline or course slideshow; topics and situations need to be reviewed in making those decisions (Worthington & Levasseur, 2015).

Acronyms: Closely associated with text but worthy of its own section is the issue of using acronyms. Acronyms are excessively used by both academia and the military with the idea that it saves time while speaking and space on a slide show, but it often causes unnecessary confusion due to lack of understanding. Even as a Sergeant Major with over 26 years of service in the Army, I am still learning new acronyms.

I have been in multiple high-level meetings were a high ranking person such as a General would be talking, using an acronym that is unfamiliar to the audience. While at a Commanders' Conference, I remember leaning over during a meeting and asking a Lieutenant Colonel, who was sitting next to me, what the acronym stood for that a General was using; he expressed that he didn't know either. So, I leaned over to my other side and asked a full Colonel what the acronym stood for. She smiled at me and replied that she wasn't sure either. Rather than interrupt the very important General, we all just sat there listening to his briefing and only partially understanding what the General was talking about. As you can see, improper use of acronyms can cause problems with communications.

Proper military writing style dictates that the first time an acronym is used it must be spelled out so that all readers know what is being discussed (Army Regulation 25-50, 2013). I have heard many people state that: "You don't need to spell out an acronym if it is one that everybody already knows; like HMMWV." My counter to this is that I believe most might know what a HMMWV is, but many don't know what it actually stands for: High Mobility Multi Wheeled Vehicle. Additionally, one runs into the common situation where we think everyone knows the acronym, but in fact, many do not. When instructing a class on the use of acronyms I often ask my class: "Who here thinks they know what the acronym COE stands for?" I typically get the majority of the class to raise their hands. I then state, "How sure are you, because you are probably wrong. By using the acronym COE do I mean Corps of Engineers, Center of Excellence, Contemporary Operating Environment, Council on Education, Contingent Owned Equipment, etc." One would never be a hundred percent sure unless it was fully written out to ensure complete understanding by all.

Always fully state what an acronym stands for. It is best to explain it both verbally and in writing. If you are giving out handouts, include a listing of acronyms and what they stand for. An additional idea is to have an acronym key at the bottom of your slide when using that acronym in your presentation. Remember that one of the keys to a successful slideshow presentation is maintaining students' focus, you can't do that if they aren't following along with what you are saying due to not knowing the acronym you are using.

Using the slideshow: Remember that the slideshow is just a tool to help the instructor/facilitator in adding a visual element to help explain, provide examples, and illustrate what is being taught. When presenting the slideshow please, for everyone's sake, to make it easier to see and to help maintain focus, maximize the screen and be in presentation mode, not edit mode. It is also important to remember that an entire class period should not be taken up by simply going over a slideshow, no matter how well it is designed. Instead, the impactful slideshow should be used to introduce/explain a subject to then transition into more hands-on learning. A prime example would be to receive a slide show as per the example dealing with throwing a hand grenade. First, the student could receive the short slideshow, ask any questions, then actually be allowed to throw a hand grenade simulator, and then finally be tested on a real hand grenade. A slideshow could still be used to help facilitate hands-on learning, but the key is that now the students would be thinking and doing instead of just receiving information. "PowerPoint is not a substitute for effective instruction and is most valued when used as a stimulus for elaboration, explanation, and discussion in classrooms" (Apperson, Laws, & Scepansky, 2016, p. 6).

VIDEO

Video is another powerful instructional tool (improves comprehension, motivation, satisfaction, interest, emotional connection, presence, connection, engagement) that should be fully considered and evaluated in that it has been repeatedly proven as effective (Anders, 2017, Miller, 2014; Rismark & Sølvberg, 2019). Video is extremely versatile in that it can be used to provide feedback (both to and from the student) and can present instructional content from many angles (even from extremely dangers areas), can visualize beyond regular human capabilities (ultraviolet, night, heat vision), can be paused, replayed, enlarged, and even slowed down. Yet video in itself isn't an educational elixir. It needs to be the right video, presented at the right time, and in the right way.

Students should be primed and informed as to what to be cognizant of while viewing. Pose questions before watching then ask them to compare their answers now to after the video, post video discussions should then occur in order to complete the learning cycle (Anders, 2015; Serrato, 2016). Additionally, be sure to show just the right amount of the video needed to best focus students' attention. This is important both for in-class video presentations as well as for full instructional videos to be watched online. Remember the concept of logically chunking informing (especially online) to avoid cognitive overload.

With regards to the optimal length of online videos, the research expresses the maximum length should be no more than 6 to 15 minutes (Berg, Brand, Grand, Kirk, & Zimmerman, 2014; Guo, Kim, & Rubin, 2014; Kaltura, 2015). The maximum video lengths listed correlate somewhat well with general maximum student prime learning durations (within 20 minutes) and generally perceived maximum student attention spans (Chaney, 2005; Sousa, 2011; Warrick, 2013). From experience and digging deeper into the research, the answer to optimal length deals more with the purpose of the video and the quality of the video itself. Self-made videos should be to the point, have adequate lighting, good-quality audio, at least periodically show the instructor (not just an unseen voice, this will also improve presence), (Anders, 2017). If the video deals with instruction then it can be longer (around 15 minutes), if the video is just an introduction, an assignment video, or something other than instruction, then it should be much shorter (around 6 minutes max). Instructional online videos should generally also include an active learning component either through intra-video questions or post video discussion thread forum engagement (Anders, 2015, 2017; Brame, 2015; Serrato, 2016).

Video also has the power to be used to address each of the learning objectives (cognitive processes) as organized by Bloom's Revised Taxonomy (Anders, 2017; Krathwohl, & Anderson, 2009). The following image presents ideas on how video can be used to help address each learning objective.

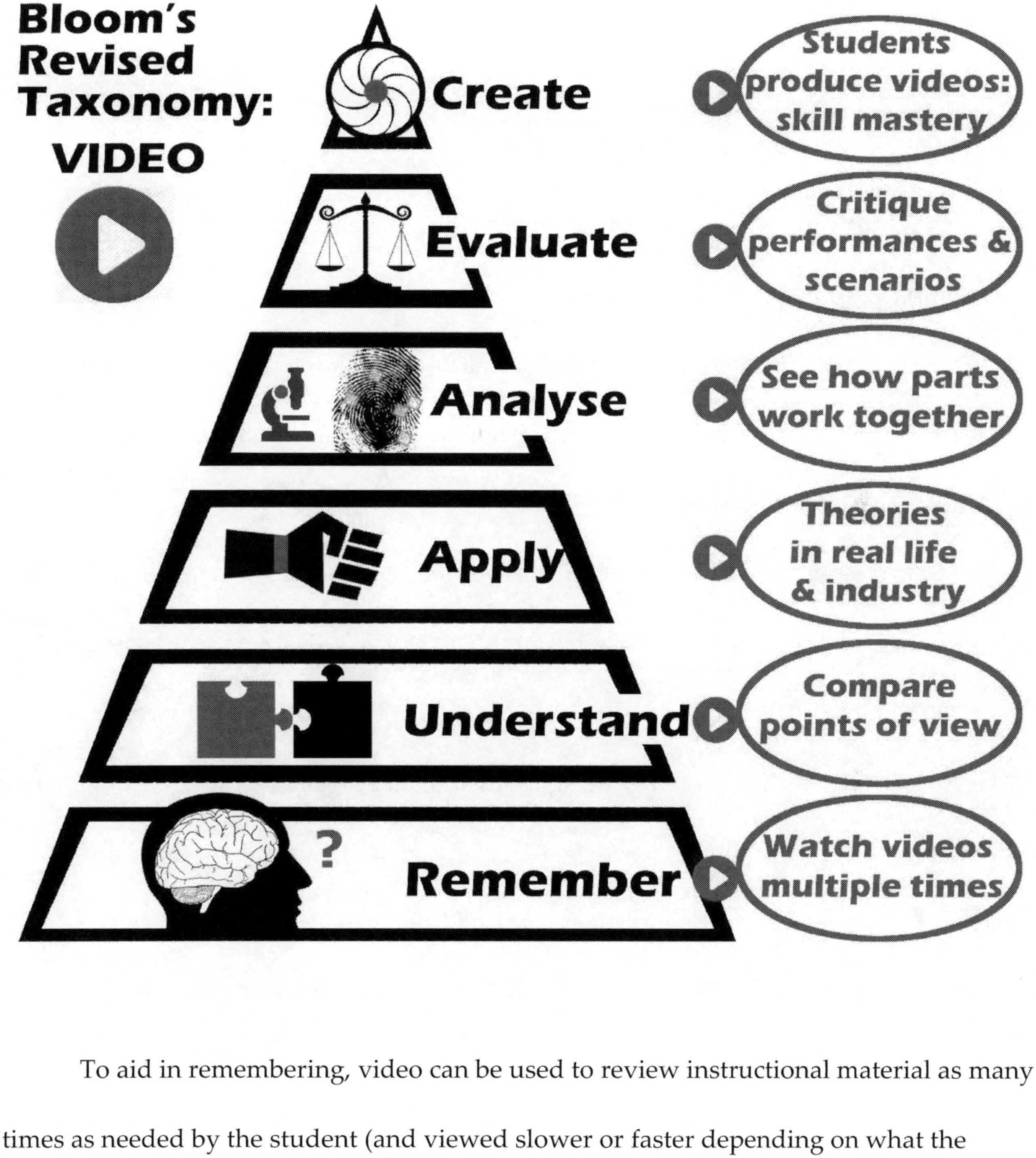

To aid in remembering, video can be used to review instructional material as many times as needed by the student (and viewed slower or faster depending on what the student desires); instructional effect is additionally enhanced when videos are closed captioned, allowing students to read the text if desired for clarification, learning issues, or environmental reasons (OFCOM, 2006; Tisdell, & Loch, 2017). Video can assist with achieving the learning objectives in all areas of Bloom's taxonomy by being able to present multiple types of information in different ways, from different perspectives, from real life, or a representation of the abstract, as well as through the power of capturing students' performances (allows for critiquing) and by permitting students to create their own videos as part of class projects.

Student video creation/production has been shown repeatedly to greatly enhance student engagement and comprehension (Campbell, Heller, & DeMara, 2019; Greene, & Crespi, 2012; Morena, Smith, & Talbert, 2019; Naqvi, 2015). Additionally, this type of hands-on learning has been noted as being motivational, fun, and memorable by students, creating a powerful educational experience. In addition to learning about the primary subject that students are creating a video about, additional skills are learned such as critical thinking, organization, video production, and digital communication skills (Greene & Crespi, 2012).

Techniques and Resources for using Instructional Technology

Instructors/facilitators must also work as instructional technology professionals in order to be able to best understand and use the right technology needed for the learning event. Use of the Analysis, Design, Development, Implementation, and Evaluation (ADDIE) model is recommended to identify and re-evaluate instructional technology needs (Training and Doctrine Pamphlet 350-70-1). An ongoing review of what instructional technologies are being used (such as online tools, slideshows, videos, etc.), how they are being used, student performance/assessment trends, ongoing student feedback (throughout the course as well as end-of-course-critiques), and mindful observations should be conducted to continually examine the effectiveness of the technologies used within a course. Costs/funding is always an issue with updating/replacing technology, but instructors/facilitators should always be monitoring for new developments with educational technologies (new technologies, new techniques, best practices, etc.).

Instructors/facilitators need to set the example by being life-long learners especially within the field of education. Learning about new instructional technology and other developments can be accomplished by reading books such as this one, going to educational conferences/presentations, following podcasts and checking web sites dedicating to educational technologies. The following sights provide good resources for news dealing with educational technologies:

- The Tech Advocate: https://www.thetechedvocate.org
- Inside Higher Ed: https://www.insidehighered.com/news/focus/technology
- Ed Tech: https://edtechmagazine.com/higher/

Additionally, the following sites provide good podcasts dealing with educational technologies:

- EdSurge: https://player.fm/series/66740
- Instructional Tech Talk: https://instructionaltechtalk.com
- Teaching in Higher Ed: https://teachinginhighered.com/episodes

SUMMARY: Interaction

This section dealt with interaction defined as students' actions or communications with the learning community and or the instructional content. Key aspects were in understanding the three main interaction types: student-to-instructor, student-to-student, and student-to-instructional content, as well as the two additional interaction types:

student-to-self and student-to-technology. Student-to-instructor interaction contained the elements of effective verbal/non-verbal communication, instructional presence (social, cognitive, teaching), the Army Experiential Learning Model, After Action Reviews, and the importance of cultural considerations. Student-to-student interaction contained the elements of etiquette/netiquette, the many aspects of group work, and group assessments. Student-to-instructional content dealt with assessment, Bloom's taxonomy, content relevance, making the content memorable (through emotional connection and fun), and making content challenging via the use of zone of proximal development.

The additional interaction types of student-to-self and student-to-technology were also presented to more fully comprehend the Army Learning Model aspect of interaction. Student-to-self interaction dealt mainly with reflection and student self-motivation, comprised of many components such as S.M.A.R.T. goals, and the importance of student resiliency. Student-to-technology was about important processes dealing with the creation and use of instructional technology like slideshows and video; techniques (ADDIE) and resources (news and podcast websites) were also presented. ALM necessitates constant evaluation of the newest learning science research to ensure that the most complete and effective instruction can be provided.

CHAPTER 4: CRITICAL THINKING

- Critical Thinking fully defined/explained along with its importance
- Operational Environment and the PMESII-PT factors
- Creative Thinking defined/explained along with its importance
- Detailed implementation example along with suggestions for incorporating critical thinking in other classes

DEFINING CRITICAL THINKING

Critical thinking is defined by the Army as "…the intellectually disciplined process of actively and skillfully conceptualizing, applying, analyzing, synthesizing, and/or evaluating information gathered from, or generated by, observation, experience, reflection, reasoning, or communication, as a guide to belief and action" (Training and Doctrine Pamphlet 525-8-2, 2017, p. 21). A useful tool in critically thinking about situations/problems are the Nine Reasoning Questions for analyzing problems, by the Thinking's Guide Library, Foundation for Critical Thinking (Elder & Paul, 2016, p. 26):

Nine Reasoning Questions for Analyzing Problems

1. What exactly is the problem?
2. The key question that emerges from the problem is…
3. My purpose in addressing the problem is…
4. Actively seek the information most relevant to the question.
5. Some important assumptions I am using in my thinking are…
6. If we solve this problem, some important implications are… If we fail to solve this problem, some important implications are…
7. The most important concepts, theories, or ideas I need in my thinking are…
8. The point(s) of view is/are as follows:
9. After reasoning through the parts of thinking above, the best solution (conclusion) to the problem is…

OPERATIONAL ENVIRONMENT - PMESII–PT

Additionally, I always recommend identifying the (contemporary) operational environment of the problem situation (this also ties in with the nine points just presented). By using the PMESII–PT factors (political, military, economic, social, information, infrastructure, physical terrain, and time) a greater understanding of the situational/cultural factors are understood to better comprehend actions/reactions of all involved. PMESII–PT factors could also be further analyzed by defining each within the civilian considerations of ASCOPE (Areas, Structures, Capabilities, Organizations, People, and Events), (Army Doctrine Reference Publication 2-0, 2012; Field Manual 6-0, 2015).

Another way to define critical thinking is the ability to fully understand and reflect upon the situation by moving beyond pure linear thinking and contemplating the issue multidimensionally to logically and creatively address the issue while also incorporating possible second and third-order effects (ramifications). To understand second and third-order effects, empathy needs to be developed to best understand all of those involved in the situation. Empathy meaning being able to understand those individuals' points of view, feelings, thoughts, and possible emotional responses (cultural understanding greatly helps with this).

CREATIVE THINKING

The Army defines *creative thinking* as a cognitive process that "involves creating something new or original; thinking in innovative ways while capitalizing on imagination, insight, and novel ideas" (Training and Doctrine Pamphlet 525-8-2, 2017, p. 21). Both critical and creative thinking are essential for the Army planning methodologies of the Troop Leading Procedures (TLP), the Military Decision-Making Process (MDMP), and the Army design methodology (Army Doctrine Publication, 5-0, 2012), as well as for strategic thinking. Army leaders have also stated that: "individuals, groups, and organizations at the tactical, operational, and strategic levels need to be creative to provide new and effective approaches to challenges and understand the interaction between an organization and its external environment" (Allen & Gerras, 2009). Due to the importance of creativity, it should also be developed within ongoing instruction.

Two great ways to help students improve their creative thinking skills is by reviewing case studies and by having learners brainstorm ideas. Having students review novel solutions to real-world issues via case studies provides them examples and interesting ideas to contemplate. Be sure to discuss the characteristics of the ideas within the case studies and how they might have reached that solution. Another creativity-enhancing technique is brainstorming. This active process forces students (either as a group or individually) to come up with different ideas and then create even more ideas by combining previously formed notions. Ensure interactive discussion and even constructive critiquing of ideas to generate even more creative thought (Feinberg & Nemeth, 2008). "Practicing creative and critical thinking in a non-threatening schoolhouse or training environment enhances problem-framing (design) and problem-solving (planning) skills" (Training and Doctrine Pamphlet 525-8-2, 2017, p. 21).

IMPLEMENTATION EXAMPLE: Critical Thinking

For this topic, the implementation example will serve as an effective way to best describe the aspect of critical thinking within instruction. In the following example, I pose a problem and then work through the issue with the class through the use of critical thinking. In this way, we learn to use a better way of thinking by incorporating critical and creative thought processes to best solve a presented issue.

"Welcome class, here is the situation. A Middle Eastern country that we are working on building better relations with has reached out to us, asking for help. A remote village (small town) has been experiencing severe drought. People are starving to death.

This country has asked for assistance from the U.S. specifically. In classic American style, we address the situation as quickly and effectively as possible. It is a pretty simple situation, right? There are people that are starving so we will give them food and end their starvation. So, the Army quickly sends a unit to the area, sets up local security, and distributes a large amount of food - boxes of Army meals ready to eat (MREs) to the population. Now, what is the end result? The result is anger and attacks by the local population, with our soldiers getting hurt. Why do you think this happened? "

At this point, I pause and wait for students to contemplate and then give their reactions. I usually ask a few more students for their thoughts as well. After a bit of discussion dealing with their thoughts, I put forth this answer:

"The reason why the people responded negatively is that we (the United States) employed linear thinking instead of critical thinking. We didn't fully think of the second and third-order effects of how we addressed the issue. "

Together, as a class, we come up with good definitions for linear thinking and critical thinking, and then compare that with *what the book says*.

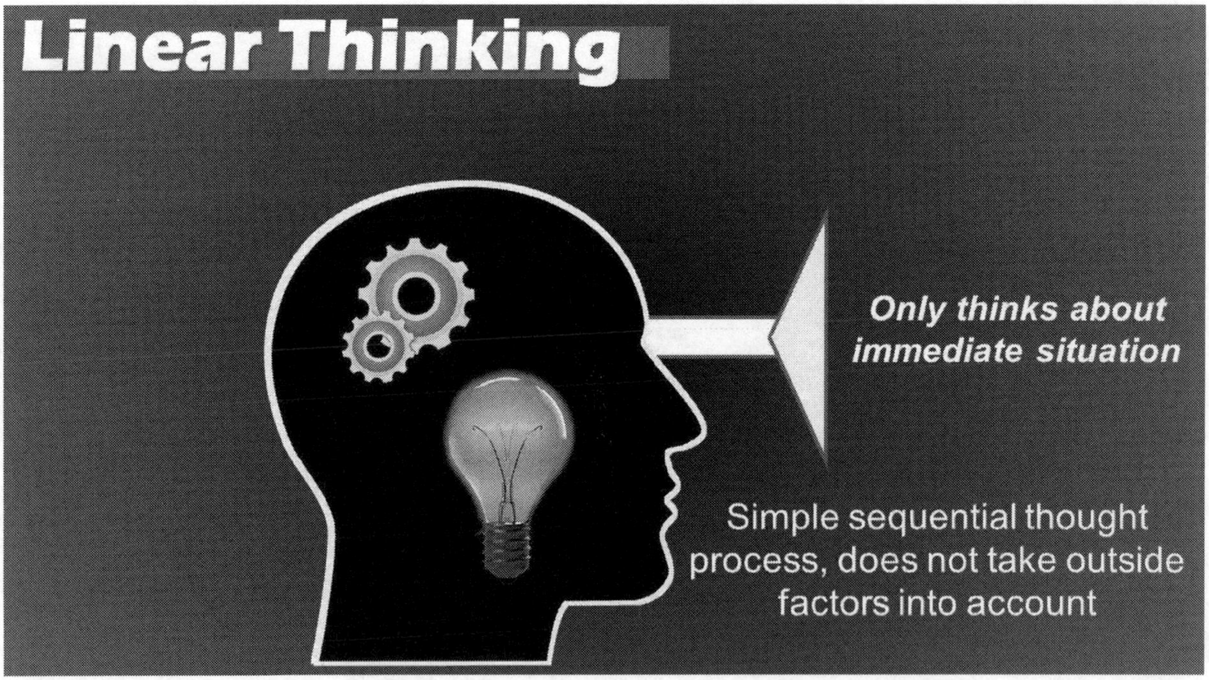

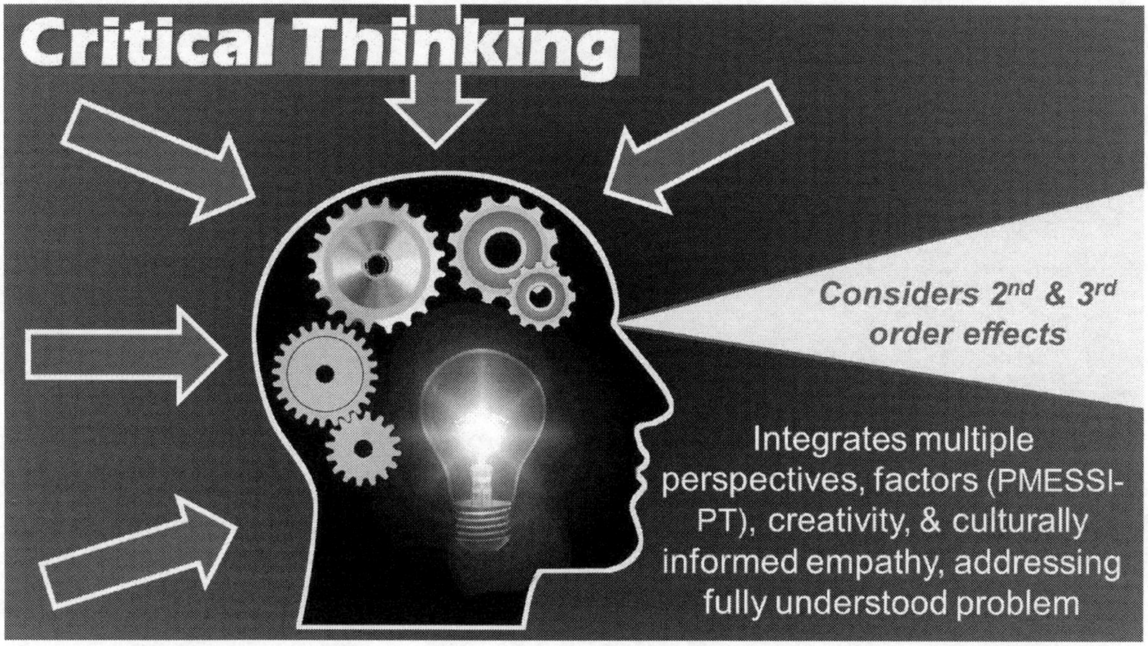

"To avoid the previous negative results lets restart this scenario. Let's reflect on what happened when we used linear thinking and seek ways to avoid that result. This time lets address the problem using critical thinking. Together let's analyze this real-world type issue by using the 9 reasoning questions for analyzing problems, as well as the PMESII–PT factors."

At this point, we go into a discussion (with full class participation) to try and fully identify the problem. We use a whiteboard or flip chart to develop answers as a group to the nine reasoning questions as well as to fully address the PMESII–PT factors of political, military, economic, social, information, infrastructure, physical terrain, and time. I only provide specifics if they ask me, like "What country specifically is asking us for assistance, where exactly is this village?" If I see that someone isn't fully participating, I ask them a question directly such as "SGT Smith, I remember you mentioning in a previous class that you had been deployed to Afghanistan. What are your thoughts on the I: Information variable for this?" Throughout the class we would use subject matter experts (soldiers previously deployed to the area reflecting on their experience), the CIA world factbook, Army manuals, or other reputable Internet sources to find all the information needed for our critical thinking analysis. We would also allow for debate if someone disagreed and would work to find additional sources of information to be able to come to a consensus.

Once we come up with a better understanding of the problem and take the additional factors and variables into account, we then brainstorm and use our creativity to create a new course of action as to how we, as a group would address the problem. Once all of that is put together, I ask if one or two students feel like they fully understand the entire issue and can brief us all (the entire class) to put the whole thing together. I make sure to only step in and help if they truly need it so that the class, as a whole, doesn't get confused.

Student wrap up presentation: "We identified that the problem was that villagers in the Samangan province of Afghanistan needed U.S. assistance to address inadequate food and water supplies due to severe drought in the area. We realized that we needed to fully understand the area of operations and not simply make assumptions about the region. We have written out all of the PMESII–PT factors for this area on the whiteboard. "

P: Political: Islamic republic, President, a council of ministers, provincial governors, village leaders

M: Military: local police in the region with some military checkpoints

E: Economic: poor economy, lots of corruption, large illegal drug market

S: Social: culturally different than the United States. Strictly follow conservative Islamic religion, very male lead

I: Information: currently low literacy rate (31%), information mainly distributed through word of mouth and video via social media networks

I: Infrastructure: poor development: problems with roads, water, electricity access

P: Physical Terrain: difficult terrain: mountains and deserts

T: Time: starvation situation is occurring during the holy month of Ramadan (religious holiday): people fast throughout the day

"Based on this information we have decided that our course of action will be to work through the village tribal elders, along with the local police/military to establish a secure distribution area. We will be sure to have female soldiers in case any female villagers need to be searched in order to ensure security without violating any religious laws or cultural sensitivities. Information about the distribution site will be disseminated through village leaders, pictorial posters, and descriptive videos through social media. "

"Some additional key factors in our course of action are that food and water would be given out after sundown when the villagers can break their daily religious fast. Additionally, we would only distribute special Halal certified MRE meals, or at the very least meals that do not contain any pork in order to address the local population's special dietary requirements in accordance with their religious laws and culture. We would also bring multiple translators familiar with the customs of the area to address any issues and ensure a smooth operation. "

At this point I would compliment the class on doing a great job and state:

"Yes, thanks to your hard work in using critical thinking before your course of action, the end result is much different, much better. We have succeeded in our mission, did not suffer casualties, and gained a valuable ally in the region, all by being creative in our use of critical thinking. More advanced implementations of our course of action development would have involved going through the entire military decision making process (MDMP). What we just experienced as a class is important to remember because in today's complex world, critical thinking will be vital for all of our missions both within the Army and in civilian life."

Notice how previous elements dealing with motivation (attention-getting/maintaining, relevancy, confidence, and satisfaction) and interaction (group work, communication, instructional presence) are all still addressed. Interesting, impactful photos are used to help make an appropriate emotional connection to the information and to aid in long term retention as well as to make the class more fun and engaging. Additionally, the photo slides help make the modern scenario that much more realistic. I have typically followed up this class with an additional *practical exercise* where I use another real-world scenario but take a further step back to allow the students to address the issue more on their own (example provided in the *Interaction* implementation section).

SUGGESTIONS for Incorporation in other Classes

From the implementation example provided, a key aspect of dealing with incorporating critical thinking into other classes is ensuring that students fully think about the problem at hand. Students must move away from short-sighted linear thought processes to an understanding of second and third-order effects. Students need an understanding beyond the surface level by incorporating multiple perspectives and considering the operational environment in which the problem is occurring. Additionally, ensure students reflect on the problem and the information presented (is it like other situation, how is it different, does it correlate with their previous experiences).

Creativity should also be highlighted to let students brainstorm and come up with many different ideas to address the issue at hand (sometimes coming up with bad ideas leads to coming up with good ideas). “Creative thinking is a cognitive process that supports divergent and convergent aspects of problem solving and decision making” (Allen, & Gerras, 2009, p. 78). Also, allow for debates to occur, question the data, understand different points of view (Feinberg, & Nemeth, 2008). Finally, work to help students develop culturally informed empathy to enable a greater understanding of others’ thoughts, emotions, and actions (Facione, 2011; Training and Doctrine Pamphlet 350-70-14, 2015). This is highlighted in The Applied Critical Thinking Handbook, (formerly known as the Red Team Handbook) in describing the importance of culturally informed empathy, expressing that it

> …can improve communications with others so as to endear and not offend, to facilitate collaboration and compromise, and to settle disputes peacefully when preferable. This involves language beyond the verbal, and into customs, prejudices,

habits, mores, expectations, fears, historical grievances, community pride and the like. (p. 36).

The Army Learning Model holds the use and development of critical thinking skills as a main objective in its practice; instructors/facilitators should always look for way to further develop this capability in their students.

SUMMARY: Critical Thinking

This section dealt with defining and implementing critical thinking. Multiple definitions of critical thinking were presented, focusing on its reflective nature, incorporation of outside factors, and the use of culturally informed empathy. The Nine Reasoning Questions where presented to assist students in being able to better analyze any given problem. Similarly, the PMESII-PT factors were presented as a way to help students fully understand the operating environment of the problem. The importance of creative thinking was also explained, describing its benefit in developing novel ideas and solutions to problems.

A critical thinking implementation example was presented where a real-world type problem was presented and linear thinking was used, ending in a negative result.

Critical thinking was then used by the class as they worked to fully understand the problem and the many different factors involved. Creative thinking was then used to come up with the best solution to address the identified problem. From the implementation example, suggestions were presented dealing with how to use critical thinking in other classes to continually help the student improve via advanced use of the Army Learning Model. A final note dealing with critical thinking and something to always keep in mind is "…key to teaching critical thinking skills is to ensure the organizational climate values critical and innovative thinking. …an organization demonstrates its commitment to these values by expecting and rewarding critical thinking and innovation" (Griffin & McClary, 2005 p. 111). Instructors/facilitators must always work to value and recognize the importance of the community of inquiry in general, and critical thinking in particular.

CHAPTER 5: EXPERIENCIAL LEARNING

- Instructional Methodologies defined/explained along with its importance
- Descriptions/examples of the instructional methodologies: Case Study, Demonstration, Discussion, Games, Lecture, Practical Exercise, Problem Based Learning, Storytelling/Presentation, Role-Playing, and Simulation
- Simple formula to use when giving instruction presented: SEEI

Experiential Learning is a *process* of learning through active engagement with the learning subject (a real event/experience) followed by purposeful reflection on what occurred (fully understanding/grasping all that happened), then conceptualizing and thinking about the event (integrating with prior knowledge), proceed by active experimentation – practice (gaining skill mastery of what was learned), finally applying what was learned to a new experience (and then repeating the process, a continuous spiral of learning), (Kolb, & Kolb, 2017). It isn't enough to simply have an experience, the full process described must occur for learning to fully take hold. Experiential Learning encompasses all four modes of Kolb's experiential learning cycle and is "applicable in all learning situations both in the classroom and in life" (Kolb, & Kolb, 2017, p. 14). This can then be extrapolated and applied within the framework of the Army Experiential Learning Model (TRADOC Pamphlet 350-70-7, 2018).

Now the question arises as to how we physically and effectively apply the Experiential Learning process within our courseware/training both within the military and in higher education or business training environments. The answer lies in properly using the right instructional methodology to best address all aspects of experiential learning. There are many different types of instructional methodologies that instructors/facilitators can choose from. Yet care should be taken in identifying the best instructional methodology to use based on the topic, time/resources available, the learning objective/level of understanding, and the audience composition (new or advanced students). A brief definition will be given for the proceeding main educational methodologies with much greater follow-on information for each, presented after the initial description.

Instructional Methodologies

Case Study: students presented details of real-world problem/situation and asked to address the issue. Afterwards, students provided with real-world solutions for comparison.

Demonstration: skill/process to be learned is presented at full speed, then step-by-step. Students often expected to be able to perform the demonstration by the completion of instruction.

Discussion: class conversation where questions are asked and thoughts/ideas are shared. Instructors/facilitators focus and guide the interactions to fully cover the topic being addressed.

Games: active learning process with components of fun and competition. Various implementations possible that actively test student knowledge/skills in an entertaining manner.

Lecture: long one-way oral presentation (generally by a subject matter expert) to a group of students. Passive learning, no interaction. Often used in large classes, least liked by students.

Practical Exercise: hands-on action generally performed after initial instruction; could fall within the realm of low-stakes testing (although not graded), allows for skills mastery.

Problem Based Learning: Students (generally in groups) presented with a multifaceted problem to solve and present. More advanced method for advanced students, more self-directed.

Role-Playing: students act out different situations by assuming different positions (roles). This provides students different experiences with varied perspectives and emotions.

Simulation: students are immersed within a realistic situation dealing with the learning topic. Students successfully maneuver through the active situation to demonstrate competency.

Storytelling/Presentation: shorter than a lecture (about 20 min.) with greater impact. Purposefully dynamic & interesting, usually with media, information weaved into a story.

Case Study

This method of instruction presents a "real world" situation (generally a problem) that has already occurred (a case study). Details are provided allowing students to form a full understanding of the different people and circumstances involved. Usually, but not always, after students have received the case study (provided as a document, a video, a presentation, etc.) the instructor/facilitator talks about the case study with the students. Students are then asked a series of questions about the specific issue and ways in which the problem can be solved. This can be done in multiple ways such as by questioning individual students, the whole class, or as an assignment. Afterward, students are provided with the real-world solution for comparison against the class or individual students' solutions/ideas. Generally, this method of instruction is used to help students achieve three things: become better at understanding presented problems, find a feasible solution, and identify ways in which the problem presented could have been prevented (all within the framework of the subject being taught), (Nilson, 2010).

Case study method of instruction has been found to be most beneficial in integrating theory into practice, identify the need to use critical thinking skills, being able to fully identify a problem, and being able to make final decisions and fully defend them (Farashahi & Tajeddin, 2018; Raju & Sankar, 1999; Wlodkowski, 2008). The case study method has been heavily used and researched in the field of engineering education where they have identified that this method of instruction piques students' interest since it is a *real-world* problem, adding immediate relevancy to the situation (Raju & Sankar, 1999).

Case Study Implementation Suggestions

Whenever possible, case studies should be presented with multiple possibilities for solutions, this forces students to weigh the advantages and disadvantages to their ideas (Raju & Sankar, 1999). Full character information of the people involved in the problem should be provided. This allows students to role-play those positions to try and gain insights as to why the problem occurred and the best solution (Raju & Sankar, 1999). Additionally, do not underestimate the importance and motivational enhancement properties of using multimedia (video/images) and actual site visits (when possible) to make the subject matter come alive and become even more real to the students.

Demonstration

In this method of instruction an action is performed (generally a kinesthetic movement but not always), usually at full normal speed (referred to as *combat speed* in the Army) to approximate the real-life action that the demonstration is trying to highlight. Prime examples would be evaluating a casualty, Army combatives (hand-to-hand combat), and how to fix a malfunction on a rifle (SPORTS: Slap, Pull, Observe, Release, Tap, Squeeze). Demonstrations can also show students reactions and situations such as what occurs to an engine block when an incendiary (thermite) grenade is placed on top of it, or the effects of CS (chlorobenzylidene malononitrile) riot control gas. If the demonstration is highlighting what students are to be able to perform themselves by the end of instruction, the demonstration is generally conducted again, much slower, step by step, to ensure that students understand the movements and process (Inuwa, Abdullah, & Hassan, 2018). It is important to complete a demonstration and ensure all see the final result/process so that students understand what "right" looks like.

The U.S. Army and military in general heavily uses the demonstration method of instruction (often referred as demonstration based training) in that it has been extensively researched and proven most effective (when conducted in the right way) for the acquisition of procedural (how to do something), and strategic (when to do something) knowledge (Salas, Rosen, Pavlas, Jensen, Fu, Ramachandran, & Lampton, 2009; Taylor, Russ-Eft, & Chan, 2005). Benefits of this hands-on method of instruction include improved achievement scores, time-saving, better understanding and knowledge of real-world application, increased motivation, long-term retention, and an enhanced educational experience (Daluba, 2013; Inuwa, Abdullah, & Hassan, 2018; Salas, et al., 2009). Additionally, demonstrations have a power of their own in that they can be very exciting and are often described by students as being captivating, motivating, interesting, and fun (Bodner, 2001; Shmaefsky, 2005).

Demonstration Implementation Suggestions

Army connected research (Salas, et al. 2009) yielded a multiple of important guidelines to help ensure effectiveness when using the demonstration method of instruction. What follows is an extrapolation of their suggestions with additional comments based on my experience in the field of instruction using demonstrations in both the Army and higher academia.

Guideline 1: Key aspects (movements/process) that are to be learned, must be fully perceptible by the students watching the demonstration. If students can't see or understand that something is happening, then they will not be able to learn it from the demonstration.

Guideline 2: The demonstration being performed must be within the students' level of understanding or skill range in order to properly comprehend the event. Simpler demonstrations or only demonstrating one portion of an event may be needed if the students are not at the appropriate cognitive or physical skill level.

Guideline 3: Focus students' attention on important aspects of the demonstration. Prepare students before the demonstration by telling them what to expect and what to be looking for (key points important in being able to perform the event themselves) to get the most out of the event. This can be even more effective by having students rewrite these key points in their own words.

Guideline 4: Make covert or hard to perceive portions of the demonstration more overt/understandable by using an instructional narrative. Stating what is going on during a demonstration can help students fully comprehend what they are observing. *Example demonstration of conducting SPORTS, immediate action to fix a nonfiring rifle:* "At this point, the soldier has P, Pulled back on his rifle's charging handle and is now O, Observing to see if there is a round (bullet) stuck in the chamber."

Guideline 5: Depending on the subject matter and the students' level, effectiveness may improve by providing demonstrations on not only what *to do* but also on what *not to do*. Areas such as interpersonal skills can greatly benefit from demonstrations that offer variability to help students understand proper behavior in different types of situations.

Guideline 6: Show students the consequences of performing the demonstration both correctly and incorrectly. This helps students understand the critical aspect of correct performance as well as second and third-order effects.

Guideline7: Establish proper credibility of the individuals performing the demonstration. This increases trust and reliability to the situation, making students more willing and motivated to believe and accept the correct process being learned.

Another key aspect of the demonstration is that it is typically proceeded by a practical exercise were students who observed the demonstration must then perform it, this is typically done as soon as possible after watching the demonstration to maximize memory retention and effectiveness (Salas, et al. 2009). See the subsequent section dealing with the practical exercise method of instruction for more information.

Discussion (Conference)

This method of instruction, *discussion*, also referred to as the conferencing technique, maximizes student interaction by having all students partake in a group conversation dealing with the topic being learned. Students are asked questions throughout the conversation with the instructor/facilitator clarifying positions and ideas about the topic (as well as ensuring that the discussion remains focused and on task). Students are also given the opportunity to ask questions throughout the discussion in that this is very much meant to be a two-way communication event. The instructor/facilitator gives brief backs or paraphrases what students say to ensure that everyone is interpreting what is being said in the right way.

The conversational tone and mode of instruction help to build a greater sense of community (community of inquiry) and an environment of sharing, helping, and learning from one another (Anders, 2017). Similarly, asking for students' thoughts and opinions about the subject matter creates a greater sense of class ownership and relevancy. Research across multiple disciplines in academia has additionally found that using discussion as a method of instruction is beneficial in that it has been shown to improve analytical skills, increase critical thinking, improved reasoning capabilities, improved test scores, and creates a more favorable educational experiences (Ahmady & Ruhi, 2016; Douglas, Wells, Deschepper, & Donaldson, 2017; Murphy, Wilkinson, & Soter, 2016).

Discussion Implementation Suggestions

Discussions can be used in many ways within the classroom such as after a presentation, after watching a video, a demonstration, or after a practical exercise. The versatility of discussion allows it to be used at virtually any time. One of the many ways in which I have used discussion is by periodically implementing the Socratic method. This method is named after the great philosopher Socrates and is still heavily used in law schools and higher academia. For this technique, a series of questions are posed to the students. Depending on how students answer, new questions are posed (to guide them). Questions are continually posed until the students reach the correct answer or come to a full realization/understanding of the subject being instructed. This technique can also be used to ask probing questions of the student in order to probe their thinking and gain a deeper understanding of why an answer was given.

Discussions, in its many forms, can be used with large groups, although generally, students are more comfortable discussing in smaller groups (Apedoe, Ellefson, Schunn, 2012; Caspia, Gorskya, Chajut, 2003). Research varies on optimal group size but my experience in both face to face and online instruction leads me to believe that no more than fifteen to twenty students, either online or face to face, is the best number of students for optimal interaction during a discussion. Note that group discussion is different than group work where the optimal size based off of my experience tends to be around 5 students (again both in online and face-to-face instruction) but depends on the students and the type of assignment given (Apedoe, Ellefson, Schunn, 2012).

Using group work is an optimal way to address the issue of trying to discuss with a large group. The previously provided technique (*think pair share*) of posing a discussion topic/question and then turning it over to the class telling them to talk to "your partner on your right and left about this," or "discuss this with the people at your table for the next two minutes" is a great way to spark discussion in smaller groups and then have a table or small group representative give a response to the question posed. This then can spark additional questions and comments from a now more confident class, as a whole, allowing for better, more inclusive group discussion (Wlodkowski, 2008).

Discussion is a key instructional tool to use throughout the experiential learning process (Kolbe & Kolbe, 2017). Instructors/facilitators need to confirm that students grasp the instruction, regardless of instructional method used, by ensuring that students reflect on what occurred (discuss observations/thoughts/feelings), think about the event in abstract terms (seeing it importance and applicability to other situations), and then actively experiment through practical exercises or other applicable methods.

Remembering also that it doesn't stop there since experiential learning is an ongoing spiral system of continual learning and meaning-making (Kolbe & Kolbe, 2017).

Games

Games, as an instructional methodology can be a little hard for some people to understand because the word "games" encompasses a lot of different meanings. Within this document, games are activities that allow for freedom of failure (without major consequences), provide rapid feedback (generally via a scoring system), allow for progression, provide for some sort of fun/entertainment/challenge (and/or competition), and generally have a storyline allowing for a winner. Given the varied nature of games (physical games in the classrooms or video games on cellphones and laptops/PCs), they can be very similar to simulations, but both have some different connotations (simulations tend to be viewed as being more serious and process orientated). Simulations will be addressed specifically in a later method of instruction.

Games are beneficial for a multiple of reasons beyond the core "fun" aspect that the name itself implies. Games, in general, allow students to try something or answer something repeatedly until they get it right (low-stakes testing). By letting students play educational games repeatedly until they can get much further in the game or have beaten the game and won, this can translate into skills mastery (Federation of American Scientists, 2006; Gibson, 2006; Lin, Liu, Chen, Liou, Chang, Wu, & Yuan, 2013). Multiple benefits have been attributed to this method of instruction, with variations depending on how the game is put together, implemented, and whether it is a multiplayer or single-player game. Besides the learning of the specific focus of the content of the game, research has shown that games can help students with critical thinking, decision making, motivation, social interaction, communication, hand-eye coordination (for games using a tactile controls), and allows a student to experience a new situation with an emotional connection (Castronovo, Van Meter, & Messner, 2018; Papargyris & Poulymenakou, 2005; Van Horn, 2016).

A key opportunity that games provide is the ability to induce an emotionally laden experience (Cole, Cairns, & Gillies, 2015). That in itself should be a powerful reason to consider using games within instruction, it addresses core aspects of motivation, experiential learning, interaction (with other students and/or with the instructional content) and can invoke critical thinking, all of which make up the Army Learning Model. Remember that creating a powerful transformative educational *experience* is the goal of all instruction. Along with invoking a powerful experience, games also have the great ability to be played again and again without affecting a students' grade or creating other negative ramifications.

Army intelligence officer and former official of Mission Command Training Program, Major Van Horn expressed the critical value of games and highlighted the importance of the term *low-stakes testing* in a slightly different way "…war games are the only venue which affords the military the opportunity to wage wars and campaigns without taking lives" (Van Horn, 2016, p. 3). In some fields, the ability to learn from trial and error must be developed through other venues such as games because failing in real life can have life and death consequences. An additional benefit of educational video game implementation specifically, is that time can often be saved in that students can interact with games on there own or in class, all at the same time or whenever their specific schedule permits (Ameerbakhsha, Maharaja, Hussaina, McAdam, 2019).

For the purposes of further explaining games as a method of instruction I will describe four different implementations: 1) Creating or using a game within the physical classroom, 2) Using a highly developed popular "entertainment" video game for educational purposes, 3) Using/creating a video game that specifically addresses the subject matter being learned, 4) Using *gamification* within the class to enhance the educational experience. Each one of these types of educational game implementations is explained in the proceeding section and examples are presented for each.

Using/Creating a Video Game Specifically Addressing the Subject Matter

This type of game implementation deals with finding a specific video game that directly deals with the subject matter that is to be learned. Although specifically created educational games (often referred to as *serious games*) with big budgets are not very common, good quality games are available and the resources/capabilities needed to create your own video game are becoming easier and easier (Widder, 2017). Check out programs like FlowLab (http://flowlab.io) and PlayCode (http://polycode.org) for more information on somewhat more developed game making, or Kahoot (https://kahoot.com) for simple game creation.

The following is just a small sample of video games that are available specifically for education. Do an internet search for "your subject matter" + "video game" to see what is available and if it might work to enhance your course and instruction. In addition to using a regular search engine, a search should be conducted on the Steam (highly popular video games distribution source) website at https://store.steampowered.com.

EXAMPLES:

America's Army: Proving Grounds Game:

The official game of the U.S. Army and part of the highly acclaimed America's Army game series. Uses real weapons of the U.S. Army, teaches small unit tactical maneuvers, provides an experience (at least a little bit) of what it's like to be a U.S. Army soldier and focuses on teamwork to achieve victory. This is the rare educational game that also has a multimillion-dollar budget, it also serves as a recruiting tool.

https://www.americasarmy.com

The Uber Game:

You play the role of a full-time Uber driver. Based on statistics & interviews with real Uber drivers. This game provides an emotional understanding and experience of what it is like to be an Uber driver, while at the same time teaching communication and budgeting skills.

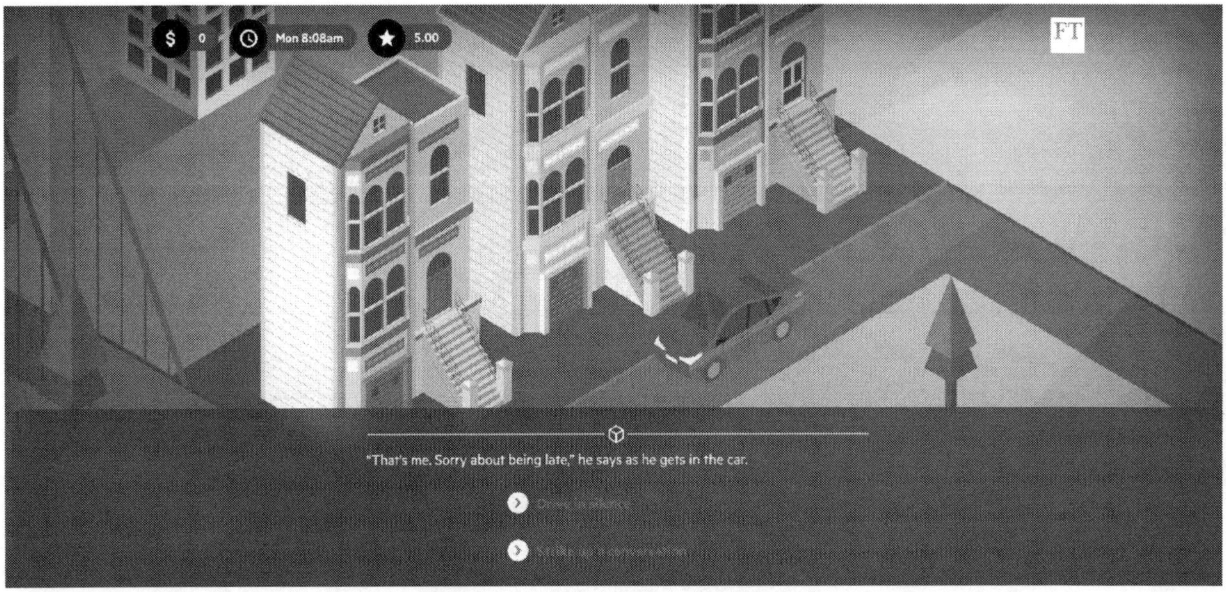

https://ig.ft.com/uber-game/

Rescue Medic:

Experience what it's like to have to make quick decisions in deciding how to triage a group of people who have just been in a major accident. Then work to save each of their lives.

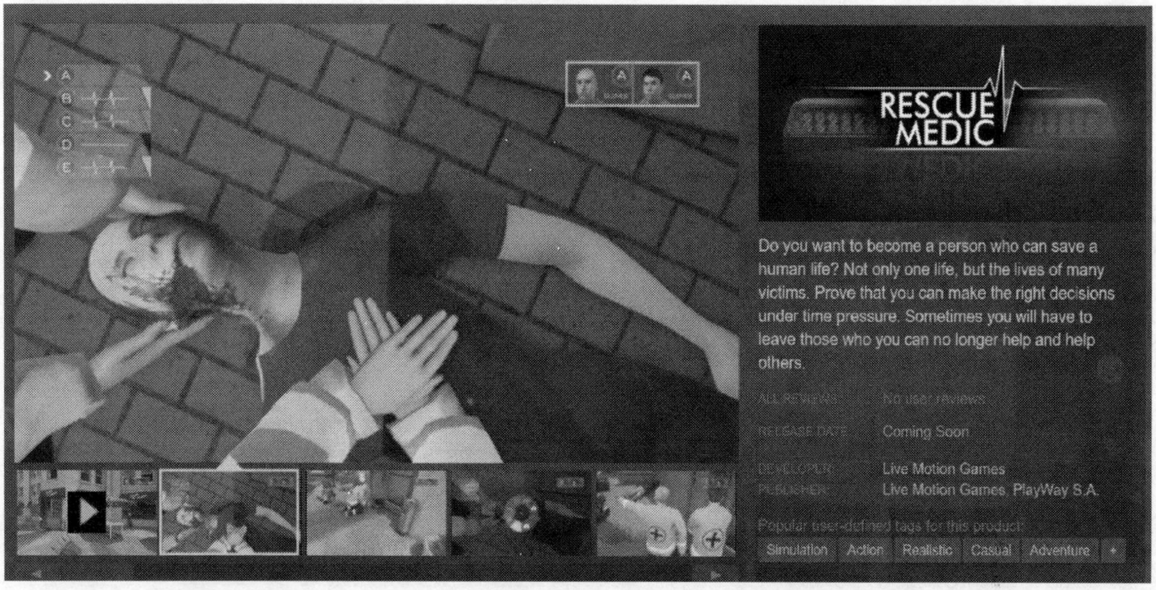

https://store.steampowered.com/app/942220/Rescue_Medic

American Mall Game:

A fun simulation-type game about the real struggles of brick-and-mortar American malls around the country. Players choose a character and then have to make hard business decisions (math, critical thinking, evaluations) to try and keep the mall running.

https://www.bloomberg.com/features/american-mall-game

Crossing Boundaries:

Explores the meaning of consent, peer pressure, different cultures, and thought processes through different, sometimes emotional situations. Interesting scenarios presented where players use critical thinking, communication, and reflection. Very story-driven with emotional connection.

https://jag.itch.io/crossing-boundaries

Plantville Interactive Game:

Provides a near-real experience (stress and feelings) of being a plant manager while working to improve productivity, efficiency, sustainability and the overall health of the facility. Skills used/learned include math, quick thinking, planning, communication, and general management.

https://www.automation.com/library/multimedia-library/plantville-interactive-game-by-siemens

Creating or Using a Game within the Physical Classroom

This is more of the classical sense of games for instructional purposes. This could be something as simple as breaking the class into different teams and seeing who can properly accomplish a task the fastest or splitting the class into two groups and then asking each group a series of questions and seeing which group gets the most answers correct. This would provide a challenge, be fun, provide a score, and would acknowledge a winner. This can also become more complex, fun, and impactful, through the use of visual/audio aids such as a large timer, a leader board, a scoring system, or even through the use of a Jeopardy® gameshow-like slideshow template, where students would answer questions as if they were on the game show.

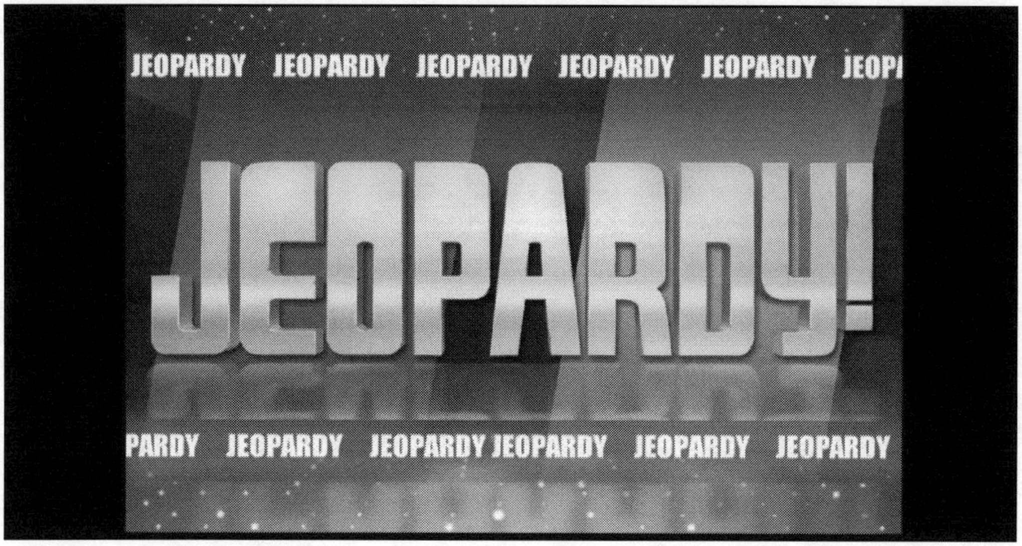

Free PowerPoint template with sound at:
https://www.youthdownloads.com/games/jeopardy-powerpoint-template.

Another free version available at:
https://docs.google.com/presentation/d/1N_5IbXUY3y2PCuhFQ0YA7ZuREwC7ew1Q3fyILBnEBQA/edit#slide=id.p.

These are not my PowerPoint links and I am not associated with them in any way; they are just a resource recommendation. If the links are no longer valid, do an Internet search for "free PowerPoint template for Jeopardy". For variations, you could also try "free PowerPoint template for Wheel of Fortune" and free PowerPoint template for Who Wants to be a Millionaire." Note that Jeopardy, Wheel of Fortune, and Who Wants to be a Millionaire are registered trademarks and I am also not associated with those companies.

Research has been done on this specific technique and implementation (using a Jeopardy gameshow-like PowerPoint in higher education) and it has been proven to be effective (Aji, & Khan, 2019; Kadoun, 2015; Shiroma, Massa, & Alarcon, 2011). But considerations should always be made when implementing a game like this in the classroom. What is the applicability and suitability of this technique (what type of audience/students do you have, what is their skill level, how much time is allowed, the subject matter, etc.); be sure to contemplate each one of these issues before implementation.

I have used this technique (using the Jeopardy gameshow-like PowerPoint template) before in a course that prepared students how to be better instructors (*Army Basic Instructors Course,* now called *Common Faculty Development-Instructor Course*) and the students (aged from 22 to 50) expressed that they enjoyed the activity very much. All students were actively engaged, smiling, and were having fun. For added enjoyment, I even had a prize table with different prizes such as bottles of water with a personalized label (it was just a paper I printed and cut out that had the name of the course plus the phrase "Education Water" on it, students loved it and found it hilarious). The game was used twice within an 80-hour course to review the information (check on learning) after the subject matter had already been presented via another instructional method. This was done to keep the game/technique novel and fun in and of itself.

There are many other ways to incorporate games into the physical classroom. Using board games such as Taboo®, or Clue®, other games like charades, 20 questions, or even escape room games, but with your own questions or rules, have also been used with success in higher education (Cervantes, 2009; Harris, 2014; Zhang, Lee, Rodriguez, Rudner, Chan, & Papanagnou, 2018). Search the Internet and use your own or your students' creativity in using games to incorporate into the physical classroom. Start with a small, short-duration game implementation and see how that goes and then build up from there.

There are some things to be careful of when implementing games in the classroom. I have already stated that you need to be sure that the game is suitable for your audience/students; some games could be perceived as too elementary (simple or childlike). Be sure to explain why you are using the game and continually evaluate the game to make sure it is appropriate. Sometimes players (students) may say it is too elementary before they play the game because they are shy and then afterward express how much they enjoyed the event. Students can also have a bit of too much fun and start yelling or laughing loudly which could disrupt nearby classrooms. Wouldn't that be such a nice problem to have, students having so much fun learning that they are yelling and laughing so loud? Also, as I've stated with my experience, don't overuse the same game, keep things new and interesting to maximize motivation and curiosity. After the conclusion of the game be sure to go over the results and summarized what was learned (complete the learning process).

Using a Highly Developed Popular "Entertainment" Video Game for Education

Anther implementation for games in education would be to use a current, popular, already fully developed video game (generally referred to as an "entertainment" game) and repurpose it for educational use. The benefit of this would be that you would be using a game that has already been proven to be highly popular and playable. The difficulty would be in how to repurpose the popular entertainment game for your own instructional purposes. Thankfully there are a few different ways to do this.

One technique is to see if the game offers different settings (such as language preference) that would address what you are trying to accomplish. As an example, researchers Anderson, Reynolds, Yeh, and Huang (2008), from the National Central University of Taiwan, were able to successfully use the popular video game *America's Army* to teach Taiwanese students how to speak English. In this case, they were able to repurpose an entertainment game that was originally made with a huge budget to teaches soldier skills and repurpose it for language acquisition. In a similar way I have heard of professors in the U.S. assigning popular video games to be played outside of class as homework but set to Spanish mode to also assist in language acquisition; others have also recommended this successful technique (United Language Group, 2019).

https://www.idsoftware.com/en-us

Another excellent example of repurposing of a video game is by creating a mod (modification) for it. The best example of this was done by the U.S. Marines to one of the most popular video games at the time (mid 90's) Doom II from ID Software (Gault, 2014). The U.S. Marines at the time wanted to enhance their training dealing with increasing reaction times, being able to think fast in the face of danger, improving team movements, and communication. The fast-paced, first-person shooter Doom II video game worked perfectly for this because it also allowed for multiplayer mode and the use of a mod. Instead of playing as a space marine on the planet Mars fighting demon/monsters with special weapons like the game was intended, it was changed/modified to address a different need. The U.S. Marines were able to modify the game to use more realistic/current weapons and change the scenery (maps) to real-world areas such as U.S. Embassies, fighting against more humanoid enemies. The game was highly motivating, used throughout the Marines, and completely served its purpose. The only costs were $50 for the game and the time used by the U.S. Marines that researched and modified the game (Lt. Scott Barnett and Sgt. Dan Snyder).

Another technique would be in focusing on one portion of a game that deals with your specific subject matter. Examples would be using games such as "Angry Birds", "Fortnight", and/or "World of Warcraft" to teach velocity, economics, and math due to aspects of the game that use these components. Another example would be to use a game such as Plague, Inc.:

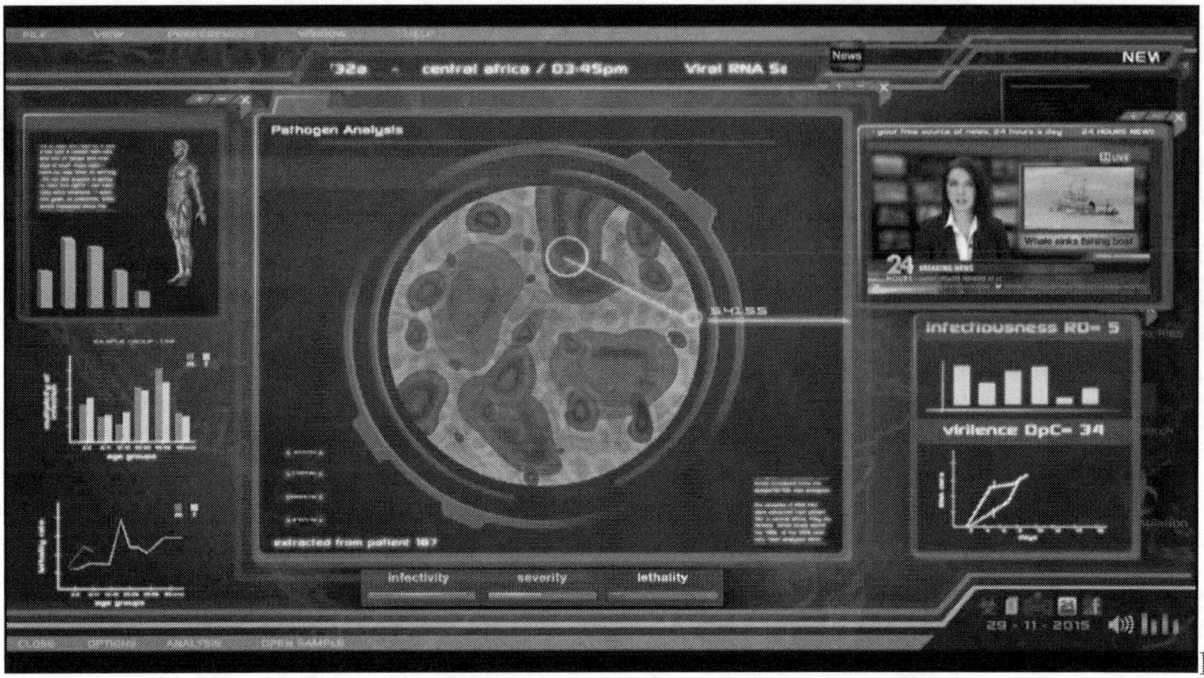

https://www.ndemiccreations.com/en/22-plague-inc

The main purpose of the Plague, Inc. game is to infect as many people as possible to win, but by refocusing the student, they can learn a lot about how diseases disperse through travel around the world, as well as the history of previous plagues, and the concept of cause and effect, along with second and third-order effects.

Again, searching and learning about currently available games and using one's creativity is the only limitation on using this type of game implementation.

Using Gamification within the Class to Enhance the Educational Experience

Although there is no unifying definition of gamification, it is best described as the deliberate implementation of game-design into a course via the use of elements such as leader boards, badges, levels, points/score, and/or presenting the course as more of an adventure/quest as opposed to a regular class (Subhash, & Cudney, 2018; Wiggins, 2016). Research on the use of gamification in higher academia has found positive results with benefits such as improved student attitude, motivation, enjoyment, engagement, and performance/achievement (Majuri, Koivisto, & Hamari, 2018; Subhash, & Cudney, 2018).

A systematic review conducted by Suhash and Cuney (2018) of 41 research articles published since 2012 revealed key elements for a successful gamification implementation. The following is an infographic followed be detailed information dealing with each gamification element.

Gamification Essentials

Points: Earn points/rewards to Level Up and win (start at zero like a real videogame)

Badges: Recognition of student-warriors' skills mastery accomplishments

Levels: Show progression through the overall mission of completing the class

Leaderboards: Display ranking to recognize – healthy competition

Missions/Quests: Special assignments for points and leveling up

Feedback: Quick and useful feedback to help the warriors continue their quest

Graphics/Imagery: Motivates & makes the game come alive

Fun: Continually update, ask for student input, & improve the course, keep it enjoyable for all!

Points: earned game score for successful mission/quest (assignment) completion, can also be used as a completely different way of grading students. Instead of a student starting with an A and losing points by not scoring perfectly on each assignment (typical technique), a student starts with zero points and has to keep going on quests to earn more points/rewards to level up and succeed.

Badges: for recognition of warrior's (student's) skills mastery accomplishments. Badges are a great way for students to physically see an accomplishment which could even be used to show others such as fellow students or potential employers.

Leaderboards: displaying student ranking to recognize and motivate to improve – healthy competition (generally done via student number instead of student name, but a decision to contemplate, using names can greatly increase competition via social pressure)

Levels: to show accomplishment, improvement, and progression through the overall mission of completing the class. Even better gamification if a level can be repeatedly "played" to gain a better "score" allows for even greater skill mastery.

Missions/Quests: the class overall is a fantastic game filled with mission/quests, not regular assignments like in a regular class (perception to promote for motivation).

Feedback: quick and useful feedback to help the warrior to see what they did wrong, cues to how to perform better to be able to continue their quest.

Additionally, the **proper use of graphics** to continually help students feel like they were part of a quest helped the gamification feel and motivated students (Suhash & Cuney, 2018). Use a cool theme for your course and be sure to remember to ensure that **the whole thing is fun**; gamification is meant to make the course more enjoyable for everyone.

There are many ways to gamify a course and the entire course doesn't have to be completely gamified. Other important tips are to remember for gamification or games in general is to still think about your overall pedagogical/andrological approach, ask for feedback/consider your students' point of view, provide different options where possible, always highlight relevancy of the content being learned, slowly try integrating games/gamification, continuously evaluate and modify as needed to suit your specific course/students' needs.

Lecture

Originally the educational term "lecture" was derived from Medieval Latin which referred to someone reading out loud and then commenting on what was read. A lecture, in its purest form, is an instructor/professor/subject matter expert in front of a group of people, talking to them (one-way communication), giving out lots of data and information for an extended period of time (often the entire class period). It is generally disliked by most people and considered boring and monotonous. The focus is more on the instructor instead of on the student, as Doyle (2011) put it, the instructor is required to: "fire his own neuron networks rather than requiring the student to build their own new networks or fire networks of prior knowledge" (p. 51). Another researcher (DiCarlo, 2009) notes that the sad reality of a lecture is that "…too often information is transferred from the notes of one person [instructor] to the notes of another person [student] without going through the minds of either person" (p. 259).

More recent higher academia research continues to show that the traditional lecture instructional technique is not as effective as more active learning modes of instruction (Horne, & Rosdahl, 2017; Russell, Andersland, Van Horne, Gikonyo, & Sloan, 2017). In her book *Teaching at its Best,* Dr. Linda Nilson cites multiple research in stating

"…experiential learning methods, such as simulations, games, and role-playing, ensure higher student motivation, more learning at higher cognitive levels, greater appreciation of the subject matter and its utility, and longer retention of the material than does the traditional lecture" (2010, p. 145). Yet the traditional lecture is still widely used.

With so much evidence against the traditional lecture, why is it still used? Why does it still exist in many places in higher academia as well as in many places in the Army? The main reason is the presumed sense of saving time. An instructor can cover a lot of material by talking non-stop for an entire class period, just lecturing the entire time. But students aren't really learning anything (Swanson, Case, Luecht, & Dillon 1996) and if the information is that important it should already be written out. In paraphrasing the words of the famous mathematician and philosopher Alfred North Whitehead, *there is no justification for someone who's sole job is to simply pass along information since the invention of the printing press in the fifteenth century*. Additionally, it takes time and planning to create meaningful learning experiences as opposed to just lecturing for an hour for a multiple-choice test to see how well people were able to memorize the data (data that will not be retained after the test), (Doyle, 2011). If the purpose of instruction is to fully learn the information, retain it, and understand how to apply it in normal and novel ways, then the use of the traditional lecture method of instruction is not the answer.

Improving the Lecture

Alternately, what I tell professors when trying to improve their educational approach is, if all you are doing is lecturing (and you feel that you must present a lecture) then you might as well record it to video and place it online with high-quality picture and sound so that students can view it as many times as they need to, can read the subtitles, and can pause and start it whenever they want. Even better would be to have that lecture video broken down into manageable chunks (6 to 15 minutes), assign the viewing of some of the videos before the class period and then use the class time to review and interactively discuss the content (this is often referred to as a flipped classroom). This would be much more student-centered and better use of the instructor's and students' *lecture* time.

Lecture Implementation Suggestions

Simply put, <u>I do not recommend the lecture, ever</u>. Given the repeated research, my experience both as an instructor and as a student, it should never be used. Review the other methods of instruction presented within this section and decide which one would work best. The very nature of the lecture: it is demotivating, lacks interaction, is passive (no critical thinking), and provides a very poor educational experience means that it does not conform in the least bit to the ideals of the Army Learning Method. It should be avoided at all costs.

Practical Exercise

A practical exercise is when a student actively performs something that has been instructed to practice a technique or process to gain greater mastery of that skill. It can be evaluated but generally falls within the realm of low-stakes testing (check on learning) and is meant as a way for students to personally experience the instruction provided. A prime example would be students receiving a class on how to use a specific computer program. Students would first see the instructor use the program, have a short discussion on the relevancy of the program and how it is used, and would then go through a practical exercise where they would actually go through and perform the learned computer program on their own (although observed by the instructor/facilitator).

I distinctly remember teaching an official Army course and realizing that students were not satisfied with the course due to too many lectures. The course did have some practical exercises, but I felt there were too few, so I created my own and doubled the total amount. My students expressed that they very much enjoyed the practical exercises yet in the course evaluations the number one thing they asked for was more practical exercises. I reviewed the course material and doubled the number of practical exercises again. Students expressed how much they loved the hands-on learning but, in the end-of-course-critiques, once again asked for more practical exercises. Why, because students love hands-on (experiential) learning.

Most students prefer hands-on practical exercise learning much more than any other instructional method (GlobeNewswire, 2014; Mosca, Curtis, Savoth, 2019; Qualters, 2002). It isn't hard to see why since people, in general, would rather do something than just sit passively as illustrated from prior experiences as an instructor and as a student myself. The end of course comments I've received in the past from students were quite direct in this observation:

> "I learn by doing so I really liked all the practical exercises."
>
> "I get antsy if I have to sit around too long. I liked how you broke things up and had us actually involved in the learning."

"You're a great presenter but my favorite times in the class were when I was up on my feet and actually doing the practical exercises."

"The hands-on learning in this course was great. I wish all courses were like this. I wish the entire class could be nothing but a series of practical exercises."

Research on the hands-on learning practical exercise method of instruction has been ongoing for over a hundred years and the results have been very positive (Dewey, 1910; Kolb, Boyatzis, & Mainemelis, 1999; McNeill & Payne, 1952; Mosca, Curtis, & Savoth, 2019). The benefits of using practical exercise as a method of instruction are that it provides a real impactful physically active educational experience, enhanced engagement, greatly increases motivation, provides greater focus, improves understanding, allows for a shared kinesthetic understanding, brings the theoretical into the real world, allows for additional discussion, and improves long term memory (Ameerbakhsh, Maharaj, Hussain, & McAdam, 2019; Dhanapal & Shan, 2013; Qualters, 2002; RAFT, 2013; Van Eynde & Spencer, 1988).

Practical Exercise Implementation Suggestions

In addition to the commonsense tips of ensuring all the equipment needed is ready to go and that all safety issues have been properly addressed for the practical exercise, there are additional good suggestions that should be taken into consideration. As in all instruction, but even more so with regards to practical exercise implementation, a positive class environment/culture must be established (Sousa, 2011). Students might fail in the practical exercise and that is OK; students need to know that it is OK and that we will learn for success and from failure (Argyris, 1991; Darabi, Arrington, & Sayilir, 2018). Ensure that the practical exercise is properly managed, all students know what is going on and why, what the criteria are for success (standards), and properly observed and assisted if necessary (AO Foundation, 2014).

Additionally, ensure proper time management; this is a vital element to ensure that all students participate and that all needed instruction/activities are covered and occur properly (Sargeant, 2012). Finally, post-exercise discussions should occur to allow for reflection, clearing up any issues, and summarizing the main points of the exercise (Di Stefano, Gino, Pisano, & Staats, 2016; Qualters, 2002).

Hands-on learning via practical exercise is basically the gold standard of experiential learning. In reflecting on my own time as an instructor, I have been tempted to do nothing but practical exercises for the entire course because of its effectiveness. Even though the practical exercise method of instruction can be a lot of fun and very motivating, and as much as some students want it all to be nothing but hands-on activities, there needs to be time taken to discuss and ensure that students reflect, realize, conceptualize, integrate, experiment, and continually apply the process of learning (Kolb & Kolb, 2017; Odo, Adenle, & Okwori, 2012).

Problem Based Learning

Problem Based Learning (PBL) method of instruction is generally considered a more advanced instructional method because it is very student-centered to the point of requiring much more self-directed learning from the student, but all students can be brought up to this level (Delisle, 1997; Major & Palmer, 2001). PBL, originally developed to improve the study of neurology, involves presenting a problem to a student (usually a group) and then having them work on their own to solve the problem through research, deliberation, integration, and finally, presentation (Barrows & Tamblyn, 1980). Generally done in small groups, students work to first fully understand the problem presented, then break down the complex issue into areas needing additional research (which is divided up amongst the small group and researched outside of class). Students then work to deliberate an agreed-upon solution for the problem and then work to create a report which would then be additionally given as an oral presentation (Major & Palmer, 2001).

Throughout this endeavor the instructor would serve only as an encourager and as a facilitator, helping only when asked and only to guide students to sources of information or to address a group structural problem (Spronken-Smith & Harland, 2009).

Although a relatively new method of instruction (developed in the 1980s), this method has been heavily used in the medical field and a fair amount within the area of business (Barrows & Tamblyn, 1980; Weimer, 2009). Reported benefits include deeper learning, greater enjoyment and fun (by both students and instructors), enhanced critical thinking and problem-solving skills, improved lifelong learning skills (self-directed learning), greater sense of ownership of learning, more positive attitudes, enhanced relevancy, and evoked greater student persistence (Major & Palmer, 2001; (Spronken-Smith & Harland, 2009; Weimer, 2009).

Problem Based Learning Implementation Suggestions

This type of instructional methodology is very learning focus and not knowledge transfer focused (students learn on their own doing research and from each other). Ensure students fully understand what is expected of them and set up any needed *learning scaffolding* such as videos on how to do properly write, how to find articles from reputable sources, fact sheets on how to avoid plagiarism, etc., ahead of time. Students need to be ready to do much more on their own in this method of instruction.

The most important aspect of this instructional method is in presenting the right type of problem to the student. To help prepare students for real life, researcher Peterson (2004) describes what types of situations work best "problems are ill structured, ambiguous, messy, complex, and most often do not have one correct answer that can be found at the end of the book in the answer key," (p. 632). This is the type of problem that students should be posed to fully develop their capabilities (Scherpereel & Bowers, 2014). I also very much like this quote because it also implies that the student can't be tested on this through a simple multiple-choice question, they must fully understand the process and be able to deliver a final product. Good examples of this would be taking an interesting business idea to address a specific need and developing into a successful business plan. Or taking a Commander's intent (to address a specific problem situation) and creating a full *operations order*.

Students need to be aware of what exactly is being assessed during this endeavor. Be sure students understand what they will be graded on and how they will be graded (use an appropriate grading rubric). One suggested assessment/rubric scheme for this type of methodology is 15% for literature review, 15% for self-reflection document, 15% for poster presentation, 15% for oral presentation, 20% for final report document, 20% for group contribution (Spronken-Smith & Harland, 2009).

The Problem Based Learning instructional methodology is a powerful tool for deep learning and student ownership of the process, but it takes time and resources (outside research time, access to Internet and possibly a library, locations for students to meet as a group, etc.). This technique also involves student presentations so adequate time management (by both the instructor/facilitator and the students) needs to be organized for educational success. Schedule and inform when each small group will be presenting to ensure proper time availability within the course of instruction.

Storytelling/Presentation

To be a good storyteller, you must first be a good presenter. Although a presentation is similar to a lecture in that it is one-way communication, it differs in that it is more like a special event with a powerful, enthusiastic presenter. To be effective it should be no longer than 15 to 20 minutes to maximize primacy and attention span (Chaney, 2005; Sousa, 2011; Warrick, 2013). The presenter needs to use expert verbal and non-verbal forms of communication (proper voice, body movements, no distractors, etc.)

with maximum instructional presence. Content presented should be elaborate, descriptive, and as impactful as possible. A *presentation* often has additional enhancing graphics (high-quality videos and/or images) that helps learners make an even greater emotional connection and grasp the information being presented. A good example of a presentation vs. a lecture would be a TED talk (www.ted.com/talks), where the preferred presentation given by a charismatic individual that delivers their information in a way that is evocative and motivates the audience, all in under 20 minutes (Tur, Harstad, & Antonakis, 2018). The most powerful presentation technique is through storytelling.

Storytelling as a method of instruction consists of presenting the subject matter to be learned by way of interweaving the material through an engaging tale/account that captivates the audience. Storytelling additionally differs from a lecture in that it requires the listener to be much more of an active learner in that "As the narrator offers their interpretation of the story the listener recreates the events for themselves through their own imagination" (Flanagan, 2015, p. 149). A good story that is presented well causes students to actively place themselves in the story (experience the story with the narrator) and/or actively compare the story to similar situations within their own lives. At the very least students actively make a judgment on the choices occurring within the story being presented. Additionally, the impact and power of the story being presented are magnified the greater the credibility of the storyteller; this implies importance in presenter introduction and delivery technique (Flanagan, 2015; Gunter, Kenny, & Junkin, 2018).

Storytelling Implementation Suggestions

The benefits of using storytelling as a method of instruction are abundant and speak to the core nature of the power of storytelling within the human psyche. Qualitative and quantitative research conducted with both students and instructors in higher academia have found that storytelling improves the practice of being reflective, induces an emotional connection, helps to reduce anxiety, allows students to better recognize the importance of culture, makes content easier to understand (makes theory more recognizable by linking it to real life), improves critical thinking, helps students make connections between the subject matter and their own prior knowledge, aids in decision making, and makes the content much more memorable (Csikar & Stefaniak, 2018; Flanagan, 2015; Tawfik, Schmidt, & Msilu, 2018).

Storytelling is most effective when the story and subject matter is seamlessly intertwined by presenting the student audience with details (the more personal the better) and a set of interesting and relatable plots and characters that must overcome a problem situation. The stories could be of examples of the subject to be learned, problems solved, illustrations of its use, analogies, etc. Effective and engaging stories tend to be ones that use lots of descriptive language to more fully immerse the audience in the reality of the story, are believable with an emotional connection, are more personal with a sense of suspense, build-up to a climactic event, and leave the audience with a positive take away or call to action of some kind.

Additional impact and enhancement can also be achieved by having students share their own relevant stories to either a small group or with the entire class (Andersson, Balslev, & Schaltz, 2019; Flanagan, 2015). This can be accomplished in several ways such as through directly asking if a student has a relevant story to contribute during a lesson, or as an assignment to be presented orally, written, by video (digital storytelling), or through other forms. Allowing students to share their stories has also shown to improve student speaking skills and overall confidence (Grange & Miller, 2018).

Role-Playing

There are many different definitions for the term *role-playing*, some of which are needlessly limiting and complicated. In reference to the Army Learning Model, role-playing is the process of having students physically act out (play the role of) a familiar or non-familiar character (role) for the purposes of interaction (Kodotchigova, 2002; Training and Doctrine Pamphlet 350-70-14, 2015). Role-play can take many forms but typically this method of instruction is used to provide the student with at least a semi-realistic experience from which to gain a deeper perceptual understanding of the subject matter.

Role-play is uniquely powerful in that it provides educational experiences in all three learning domains (cognitive, affective, psychomotor) at the same time (Maier, 2002; Rao & Stupans, 2012). As an example, imagine a student going through a class learning about how to properly evaluate/counsel a subordinate. Using the role-playing method of instruction in this situation would allow a student to learn the process of properly formulating the evaluation (cognitive), understand and properly empathize/interpret the feelings of the individual being counseled (affective), and actively use proper communication techniques (congruent verbal and body language) to physically complete the task (psychomotor). Additionally, researchers in communication in medical education expressed that "For attitude development especially that which focuses on change of affect, then role-plays should be loosely structured so that players experience emotions spontaneously. While for skills acquisition, the opportunity for repeated opportunities with feedback is critical" (Nestel & Tierney, 2007, p. 4).

The benefits of the role-playing method of instruction are that it provides a proven effective way for students to learn communication skills, helps students understand all aspects of a process (knowledge, skills, and attitudes), allows students to understand a situation/process from multiple points of view and perspectives, fosters quick thinking/problem solving and critical thinking, imparts a sense of ownership of the learning, is fun/motivating, adds action/physical movement to learning, can help reveal sticking points (misunderstandings) in students' comprehension about the subject, provides an impactful shared student experience which has a positive effect on classroom dynamics, makes the learning seem more real and relevant, also provides a memorable real educational experience with a strong emotional connection (McSharry & Jones, 2000; Rao & Stupans, 2012; Russell & Shepherd, 2010; Sutcliffe, 2002).

In addition to the powerful aspect of addressing all learning domains, role-playing is a very active (experiential) instructional method that can be a lot of fun and motivating (McSharry & Jones, 2000). I have personally used role-playing an extensive amount of time as an instructor and have also learned a lot by participating in role-playing activities as a student in many courses. Role-playing is one of my favorite methods of instruction because it is so engaging and can really benefit students by making the learning come alive in a fun and motivating way when done right. The following role-play implementation suggestions will help to ensure this method of instruction is done properly in that strong emotions can be evoked with this powerful technique.

Role-Playing Implementation Suggestions

1. **Avoid embarrassment and cultural stereotyping.** First and foremost be extra careful with this method of instruction to prevent student embarrassment and to ensure cultural awareness (Alexander & LeBaron, 2009; Nestel & Tierney, 2007). Students should never feel *ambushed* or set up to fail. College surveys across multiple institutions and departments have shown that although most students love role-playing some students are reluctant either because they have been embarrassed in the past or because they fear that they will be embarrassed (Nestel & Tierney, 2007; Stevenson & Sander, 2002). I also came to realize this early in my teaching career and developed a three-pronged approach to overcome it:
    a. **Develop the right, friendly, learning environment (community of inquiry) first.** To properly learn, students must feel safe and believe that it is ok to fail and then learn from that failure. Don't start with a difficult role-play; students need to be ready, loosened up, and feel comfortable with the class

to tackle a complex role-play. I often switch roles with students and place them at the front of the class (as a group, usually a table group of 3 to 5). I then ask them questions from the lesson we just completed as if I was a student and see how well they respond. I try to act differently each time I ask a question as if I'm a different student to vary the interaction. This really breaks the ice (relieves tension) with the group dynamics and helps the class to work together better. Part of having the right learning environment comes from telling the class (or helping them develop these guidelines) as to how students are to interact during role-play (how all are to act/behave). Students are to be respectful of each other, not *sharpshoot* (try to find fault with everything) and are to offer constructive criticism to help and not hurt. Additionally, consider any cultural/diversity issue with regards to the role-play activity. Be sure students don't fall into stereotypes because they don't properly understand the cultural/diversity issues within the role-play scenario. As an instructor/facilitator you need to properly monitor the class to ensure that students interact properly to maintain a functionally safe and friendly class environment (Clapper, 2010).

b. **Set students up for success.** A student should not have to make everything up on their own during a role-play. Be sure students have already learned some of the information and have a good idea of what is going on and what is expected of them in the role-play. Ensure the role-play scenario is at the right level of complexity and let them know what right looks like.

c. **Students need to understand and have their own resiliency**. Students need to have the right mindset and realize that if they fail or get embarrassed during a role-play event it isn't the end of the world, it is a learning opportunity. I took an advanced first aid class once and during part of the class we played different roles in treating a casualty. I was overwhelmed and didn't do as great as I wanted to which made me a bit embarrassed. I could have personally shut down, but I realized the importance/relevance of the instruction, learned from my mistake, redid the role-play, and did much better. As an instructor/facilitator you need to periodically remind students of their own strength and capabilities to overcome, adapt, and succeed.

2. **Properly warm-up the class (build up their confidence) before a full scale more complicated role-play session.** Choose a question/problem that all students can

effectively address to help orient the class to the learning subject (Jackson & Back, 2010). Have the students work in pairs to role-play different sides of a situation. An example could be that you already learned about proper information dissemination but now to leaders of the same rank/position but from different units are debating whether it is better to tell all subordinates about something at the same time or to tell junior leaders the information and then have them disseminate the information. One role-player would take one viewpoint and the other would take the other viewpoint. Having just learned this information in a previous class, students would know the reasons for each side and could have a good debate about it. This would then warm up the class for a later more advanced follow-on role-play on something associated such as leadership communication styles or how to properly correct subordinates.

3. **Create meaningful situations within the role-play scenario.** "Role playing works best when the situation involves some choice, decision, or conflict of motives. Draw situations from interpersonal conflicts, intergroup relations, moral conflicts, individual dilemmas, or historical or contemporary social problems," (Davis, 2009,

p. 160). This adds greater relevancy, realism, and more emotion to the role-play. Case-studies often have good problem sets that can be used within role-playing.

4. **Give specific set-up instructions for the role-playing scenario** (details, context, objectives) but allow students to have leeway/latitude as to how the characters are portrayed. Assign students their specific roles as well as observers (specifically tell them what to look for as well). Consider individual student skill/comprehension level but look for areas to change such as asking quieter students to play a dominating character in a scenario or having students switch roles halfway through a scenario just to give students additional perspectives and opportunities (Davis, 2009; Rao & Stupans, 2012). If there is direct assessment associated with students' actions in the role-playing, be sure to fully describe the grading rubric that will be used.
5. **Ensure the role-play lasts only as long as needed.** Role-plays should last anywhere from 5 to 10 minutes and no longer, otherwise students tend to lose focus and forget what happened towards the beginning of the event. The role-play itself doesn't have to end and is even better when it is stopped on a high point; this usually makes the students more eager to discuss, (Clapper, 2010; Davis, 2009).

6. **Give students time for reflection and then have a discussion.** Tell students after the role-play that they have a few minutes before the discussion will begin so they can use this time to reflect and jot down notes about what occurred. Classroom discussion time shouldn't focus on student mistakes, instead work to obtain all students' opinions/thoughts about the role-play and what they learned from the event (Kodotchigova, 2002). Properly guide the discussion, allows lots of student input but also pose important questions such as "what happened", "what does that mean", "how did it feel when…", "who do you think they felt when", "what could you or the other person have done differently", "what else can we do with this information," (Clapper, 2010).
7. **Be creative and consider different ways of implementing role-play**. Role-play is very flexible, can be used in a variety of ways, and can be used to focus on cognitive, affective, and psychomotor skills (or even all three at once).

Press Conference: This is one of my favorite implementations and can be done in different ways. One way is to have one or a few students act as the special VIPs and the rest of the class act as reporters. The VIPs will have done some prior reading/research about a specific case, situation, event, etc. and be ready to answer any question posed by the reporters. I like to make it even more believable by having the classroom set up like a press conference, sound effects of cameras and background noises, and making the reporters identify themselves with a made-up news agency: "Frank Smith, Kansas City Star", or "Jim Anderson, Buzzfeed."

Student Created: Look to your students for different ideas on role-play possibilities as well. Ask them if there is a subject or situation that they think would be impactful to act out. See what ideas students have as far as different ways role-play could be used for instructional purposes. This could be done as a bigger group project as well as through different medium, such as live action or via a video production.

Switching Roles: This was mentioned before but is something to really consider. Having students switch roles can be a very impactful way of teaching. As an example, I was once in a class that was focusing on how to properly conduct pat-down searches of personnel. It was one thing to play the role of the person doing the pat-down search, but it was a completely different experience being the person being patted down. This provided a completely different experience and perspective (previously I hadn't considered how the person being searched felt) that made the focus of instruction, how to conduct a pat-down search, much more well-rounded and memorable. Contemplate the value of switching roles in your implementation of the role-playing method of instruction.

Simulation

Defining simulations can be a little tricky in that they cover a broad spectrum of possibilities that range from computer simulations to physical scenario simulations as well as the use of simulators and a mixture of everything in between. Simulations are similar to games and role-playing in that they offer a realistic environment, but there are differences. Simulations involve more than just two people interacting via role-play, it generally focuses more on understanding the correct process and offers enhanced realism through environmental or objects that can be physically interacted with (such as variables in a computer simulation, a medical mannequin, or controls for a vehicle/machinery), (Russell & Shepherd, 2010). Game variables such as leader boards and scores are typically not part of the framework of a simulation (although there are such things as game simulations). Simulations can be made for any situation, but its greatest use are for situations where using real assets or environments would be too dangerous, such as life-threatening situation pertaining to the military, police, fire, and medical fields (Couwels, 2010; TRADOC Pamphlet 350-70-14, 2015). Yet simulations have been used in virtually all fields because it provides for an enhanced realistic educational experience that better prepares students for what they will face in their chosen fields (Lean, Moizer, Towler, & Abbey, 2006; Vlachopoulos, & Makri, 2017; viar360, 2017).

Newer technologies such as virtual reality allow full submersion into a simulation with 360-degree views plus up and down, with movement in all directions, interaction capabilities, and in some cases even haptic feedback, creating a true second reality. Students can interact in this virtual reality simulation either independently or together for even more interaction. Augmented reality provides an added layer of reality by superimposing data or digital artifacts into the view of the student (example you look at an engine and the augmented reality automatically labels each part for you, or you look at a foreign language document and it is instantly translated for you).

I strongly recommend all instructors/facilitators look into virtual reality (VR) and augmented reality (AR) to enhance their instruction and help students learn. Its cost of entry has gone from hundreds of thousands of dollars to mere hundreds of dollars and a lot can be done with current smartphones that virtually all students own. The educational potential of virtual and augmented reality is limitless (Gudonienė, & Rutkauskienė, 2019;

Kizil, Hancock, & Edmunds, 2001; Luckin, & Fraser, 2011). Additional VR/AR resources provided in the resource section of this book.

The benefits of using the simulation method of instruction are that it helps reduce errors by providing for multiple real life-like practice sessions, improved learning outcomes, enhanced skills mastery, increased motivation and engagement, deeper emotional connection, greatly assisted students ability in understanding theory put into practice, overall cost savings for training, and has been shown to boost knowledge retention (Krokos, Plaisant, & Varshney, 2018; Lateef, 2010; Vlachopoulos, & Makri, 2017). It can also be used to greatly reduce anxiety (as well as PTSD) by allowing students/soldiers to be pre-exposed for what they will be facing in real life, such as graphical situation, a new land/environment (Lele, 2013; Reger, Holloway, Candy, Rothbaum, Difede, Rizzo, & Gahm, 2011). Yet simulations in and of themselves aren't perfect. "A simulator does not train; it is the way the simulator is used that yields the benefit," (Marine Board, 1996, p. 67). Although this is slowly starting to change with advancements in integrated artificial intelligence (AI), it still holds true. Use of simulations/simulators needs to be appropriately facilitated with proper observations (ensuring effective use and avoiding negative learning) with time for reflection and debriefing/discussion on what was learned and how to improve (Anders, 2011).

It is also very important to understand that simulations hold a lot of power with regards to realism and inducing emotional responses from students (Gonzalez, 2019; LeBlanc, 2019). Extreme care and research should be done to determine what emotions are to be induced, why, and how this will help the student (example: prepare them for what they will experience in real life, help them overcome fear of the unknown, help them to work effectively even while under stress). As simulations can be used that are virtually life-like mindfulness should be used and continually evaluated to avoid cognitive overload and avoiding traumatization (Fraser, Huffman, Ma, Sobczak, McIlwrick, & McLaughlin, 2014; Reedy, 2015).

EXAMPLES:

U.S. Army CH-47 Chinook Helicopter Flight Simulator:

A mixture of computer simulation with physical controls to manipulate fully replicates what actual pilots in the air see, feel, and hear, allowing them to practice flying in scenarios around the world in any flying condition.

https://youtu.be/MqDrS8gjXak

Children's Hospital Los Angeles Medical Training Simulator

Full emersion virtual reality simulation that trains medical students and staff to respond to high-stakes, low-frequency pediatric emergencies. "Thanks to the immersive power of VR, we can replicate these training scenarios in true-to-life fashion, complete with paramedics rattling off symptoms, nurses and techs urging you to make a decision, and distraught parents praying for their child's survival,"

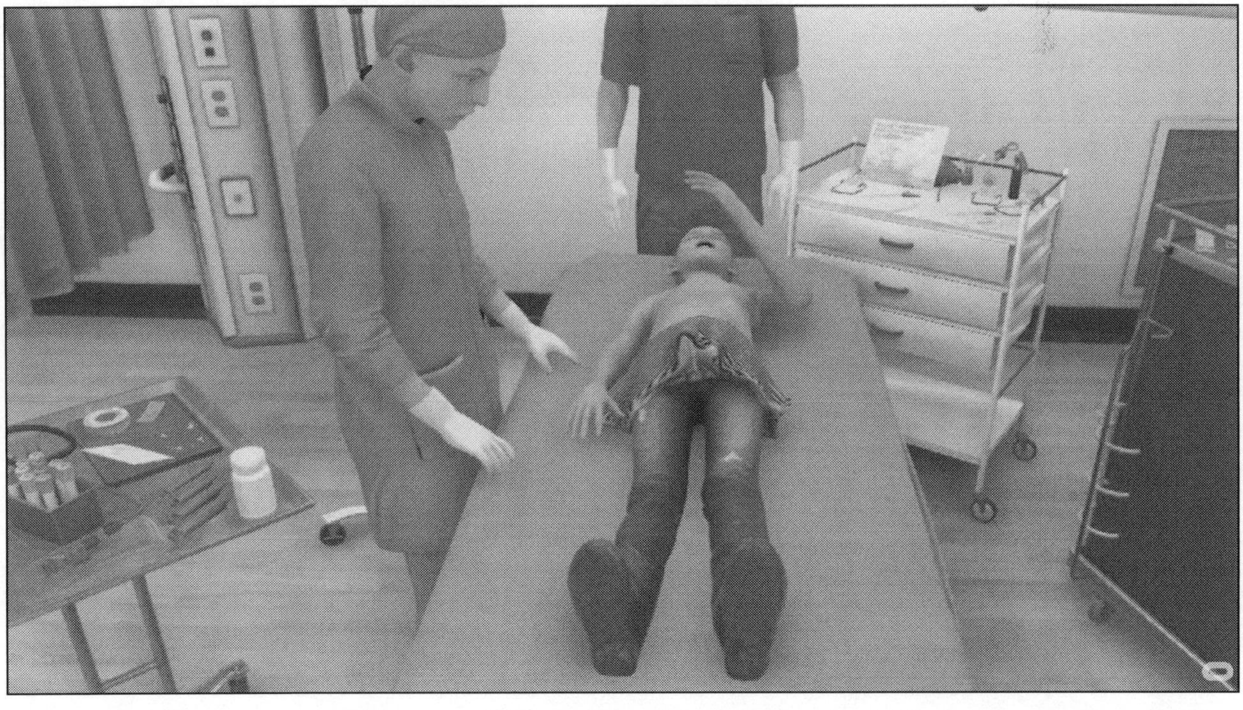

citation/video: www.oculus.com/blog/vrs-healthcare-revolution-transforming-medical-training

pH Scale Simulation:

This online interactive pH scale simulator is one of many different free simulators on the topics of Physics, Biology, Chemistry, Earth Science, and Math provided by the PhET Interactive Simulations organization from the University of Colorado.

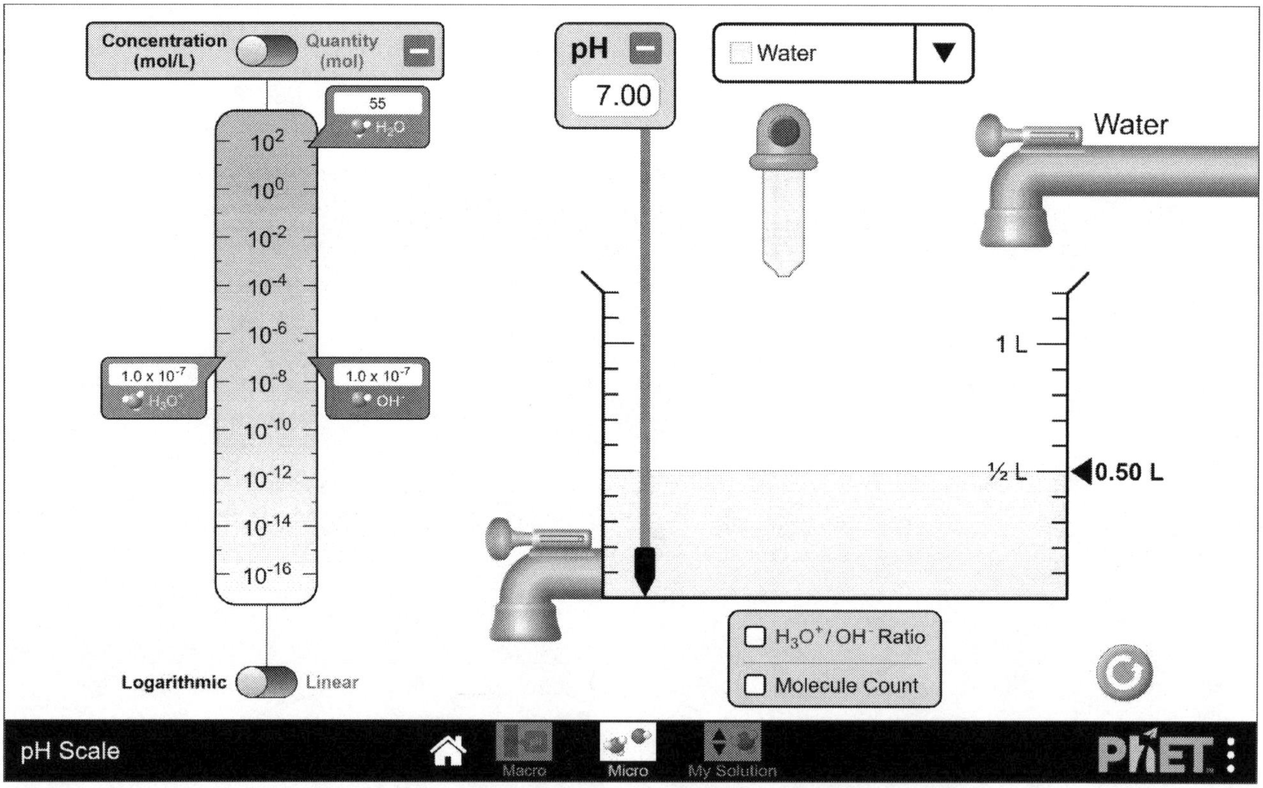

https://phet.colorado.edu/en/simulation/ph-scale

Intelligence Electronic Warfare Tactical Proficiency Trainer (IEWTPT):

This highly advanced military intelligence simulator provides scenario-driven information and data points for intelligence analysts and intelligence portions of larger military unit simulation exercises. This simulator also directly trains soldiers to build/improve questioning, information gathering techniques through avatars and multiple interactive components (AITS, 2014).

www.peostri.army.mil/intelligence-electronic-warfare-tactical-proficiency-trainer-iewtpt

Simulation Implementation Suggestions

Think safety first: A simulation and simulators are powerful techniques/tools for instruction but can be dangerous and can have second and third-order effects. Be sure all equipment is handled properly and treated as if it were the real thing (Baily, 2015). This helps to prevent accidents as well as to reduce the chances of negative learning. If a student improperly handles a weapon because it is "just a simulator" they might learn to do so when handling a real weapon. In the same way, if a student improperly handles a medical training mannequin then they might learn to incorrectly handle a real-life patient. Address these issues by establishing well-defined guidelines (before the simulation or use of a simulator), maintain proper observation, make on the spot corrections, and conduct proper debriefings.

Understand and use the relationship between fidelity and learning: Although realism should be used when students are ready, according to research conducted by the RAND Corporation, simulations don't have to have extremely high physical fidelity to help induce learning (Straus, Lewis, Connor, Eden, Boyer, Marler, Carson, Grimm, & Smigowski, 2019). This is important in that it can be very expensive to have simulations with extremely high physical fidelity, which would then limit practice on the simulation, thereby lowering effectiveness. In contrast, a simulation with lower physical fidelity, but with lower costs equaling greater access and opportunities to practice would then result in greater effectiveness for learning. Although other factors would affect this (subject matter, procedural process, end-state) it should be contemplated and evaluated. As an example, a consumer-level virtual reality headset can now be acquired for under $1000 that can provide comparable simulation experiences to simulators that cost millions of dollars (Morris, 2018).

Ensure time for reflection and discussion after a simulation: This is a good thing to do after any type of educational event, but even more so for simulations. Because a simulation can be so realistic and cause a real emotional connection, deliberate time should be taken to allow students to reflect about what occurred, take notes, and then discuss with the group their perceptions and experiences. Noted simulations training consultant Barbara Steinwachs (1992) expressed the importance of facilitating/debriefing students through three distinct phases during post-simulation discussions: *description* (let students share experiences of what happened to get a full picture from others perspectives), *analogy/analysis* (let students talk about how the simulation pertains to real-life), and *application* (ensure discussion on how what was learned can be applied to real life in different situations). This is a good recommendation and correlates similarly with the Army Experiential Learning Model. Specific individual feedback may also be needed depending on how the simulation is used (for a general experience or as part of an assessment).

SUMMARY: Instructional Methodologies

It isn't enough to simply use a good hands-on active learning method of instruction. The full process of experiential learning must be completed, students need to reflect on what occurred, they need to integrate what was learned into previous knowledge, see how it can be applied, practice the new information/skill, and then continue on in the continual spiral of learning (Kolb & Kolb, 2017). So which instructional method is best and will provide the best experiential learning? Each one of the active methods of instruction presented has their own strengths and work better for different subject matters and at different times during the class' learning process.

The instructional methods also work synergistically, meaning that by combining and using different instructional methods together, an even greater impact is made upon the students. Demonstrations work best when students have a Practical Exercise immediately following and are required to do the demonstration they just witnessed. Presentation/Storytelling is a powerful way to introduce a case-study or scenario. There are also great benefits from using Discussion after any other method of instruction (such as Role-Playing and Simulation) to help students reflect on the situation and fully grasp what occurred, as well as to discuss how to apply what was learned in new ways.

As for an implementation example, review the first implementation example that was provided at the beginning of this book (it dealt with emplacing a tourniquet). Notice how the instructor/facilitator started with a short *presentation* then introduced a guest presenter who used *storytelling*. After that there was a short *presentation,* followed by a *demonstration,* then *practical exercise,* followed by a *discussion* and then a *game*. Later, once the students were ready the next day, the instructor/facilitator increased the realism of the training by using a realistic *simulation,* where they *role-played* being in a combat situation.

Notice how the instructor/facilitator used all of these deferent methods of instruction because they were the most appropriate at the time, aided the student in understanding, and also bolstered motivation, and lead to a great educational experience. Research has proven that students gain understanding by constructing knowledge through cognitive and social connections from their educational experiences (Cross, 1990).

We as instructors/facilitators must use the latest learning science to provide high impact educational experiences to help ensure soldiers are motivated and induce long term learning.

SIMPLE FORMULA: SEEI

SEEI: State + Explain + Example + Illustration = learning

*originally *state, elaborate, exemplify, illustrate,* I modified it slightly: easier to understand/remember when presenting information.

This simple acronym was presented to me at a learning conference by famous critical thinking authority Dr. Gerald Nosich. I like it because it reminds me that no matter what type of instructional method I decide would work best, it still needs to fall within this formula to ensure that students will understand.

See the following infographic and following information for a more detailed explanation along with an example implementation:

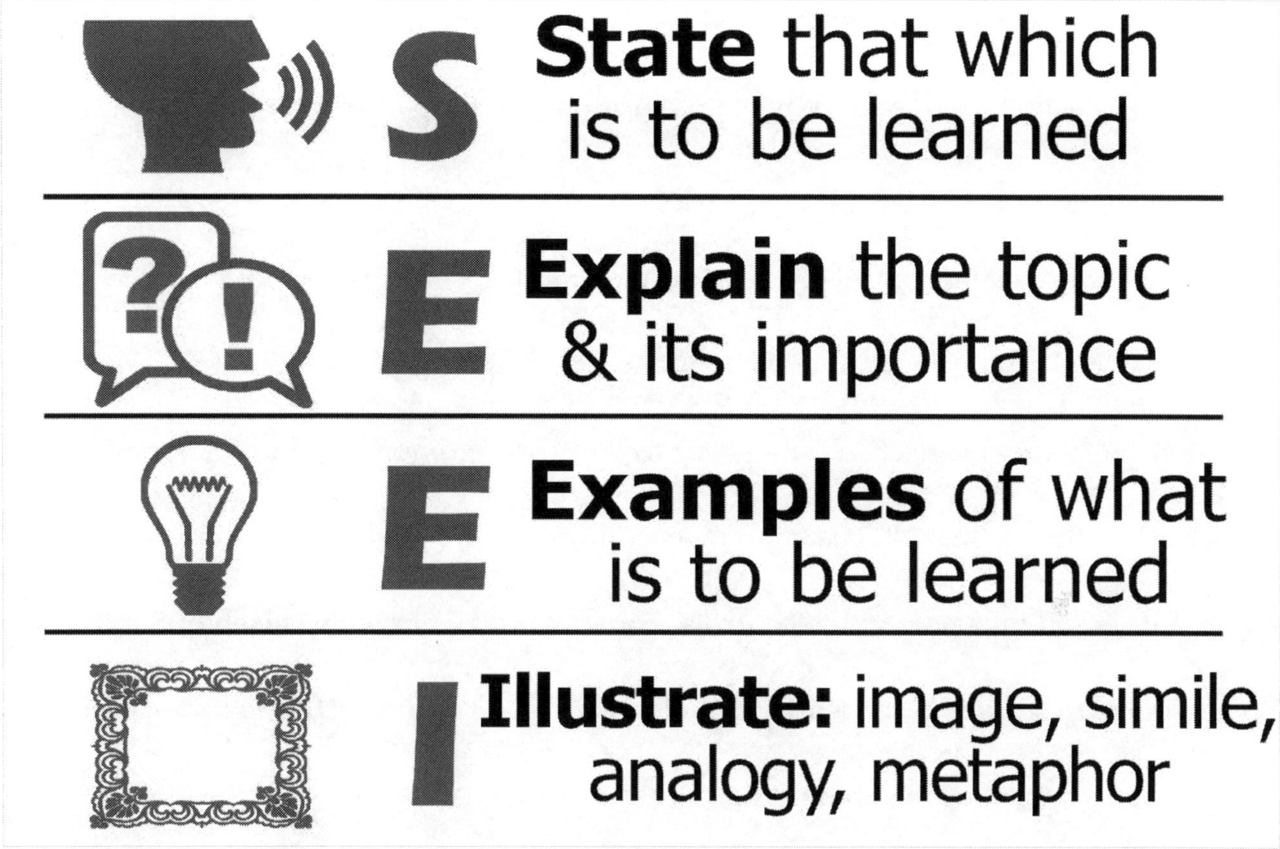

Derived from research by Nosich (2009) with slight name modification

SEEI: **S**, **State:** what it is that is being learned. **E**, **Explain:** how it works, why it is important, etc. **E**, **Example:** provide the class with examples of what is to be learned in use. **I**, **Illustrate**: provide an analogy, metaphor, or simile, as well as a picture, graph, diagram, and/or video (a visual representation), (Nosich, 2009). As an example:

"Hello class, today we are going to learn about Virtual Reality ***(S: State).*** Virtual Reality (VR) is a created digital reality that can be made into anything that one wants from abstract to realistic and can be used for entertainment as well as for education. In fact, Virtual Reality is being used by many Fortune 500 companies to improve their ability to effectively train their workforce ***(E: Explain)***. VR can be just like regular reality but much you are in greater control. To better understand and illustrate what virtual reality is, we are going to jump right in and experience it. I've given each group of two an Oculus Quest virtual reality headset. We are going to take turns going through a VR simulation trainer called Busuu Spanish Learning. Everything should be self-explanatory as soon as you put it on and take hold of the controls. You will only wear the headset for 5 minutes. We are going to see if you can learn 5 Spanish words in 5 minutes (***I: Illustrate***). After everyone gets a turn, we will have a discussion. "

Notice how each part of SEEI was addressed and a multiple of instructional methods are being employed in the example (presentation, practical exercise, simulation, discussion). No matter how advanced and complicated you get with your instructional techniques and methods, remember SEEI and you will help ensure a successful learning environment and outcome.

CHAPTER SIX: OTHER ASPECTS OF THE ARMY LEARNING CONCEPT

Another important aspect of the Army Learning Concept is the significant individual student soldier requirements. Student-soldiers must be lifelong learners and be able to both internalize and synthesize gained knowledge (Training and Doctrine Pamphlet 525-8-2, 2017). Soldiers must be *lifelong learners* (willing and able to continually learn throughout their careers), which is a 21st-century soldier skill, to be effective in their jobs. As a lifelong learner, soldiers need to maximize digital literacy and keep pace with an ever-evolving operational environment to meet current and future military engagements to include wars, humanitarian engagements, and foreign security force assistance. The battlefield continues to change at an ever-increasing rate and U.S. soldiers must be able to continually learn to stay on the cutting-edge regarding areas such as tactical techniques, equipment utilization, and digital tools manipulation. Along with being a lifelong learner, soldiers must be able to both *internalize* and *synthesize* the knowledge they obtain through formal or informal training and education (Training and Doctrine Pamphlet 525-8-2, 2011).

In internalizing knowledge, soldiers must reflect on what was learned to see how it might apply to accomplishing current and different military actions. In an associated manner, soldier synthetization of gained knowledge deals with incorporating new information with current understanding to improve one's capabilities and understanding (Training and Doctrine Pamphlet 525-8-2, 2011). Internalization and synthetization of gained knowledge are best explained through the Army Experiential Learning Model (as previously presented), (Kolb & Kolb, 2005; The Army University, 2017).

By purposefully following the Army Experiential Learning Model, soldiers can more fully internalize and synthesize gained knowledge to best achieve long-term learning with the Army Learning Concept. Long-term learning is additionally enhanced by increasing the motivational aspect of ALC through a realization of its importance and necessity (Keller, 1987). The importance and necessity of the Army Learning Concept are twofold in that the Army must be prepared to address future armed engagements and must stand ready to address any unknown issue that might arise. Future military conflicts against peer, near-peer, and unknown adversaries will require Army training and an educational learning organization that can innovate and adjust even faster than the previous decade. "Future training and education success depends on the Army's imagination and willingness to adapt" (Training and Doctrine Pamphlet 525-8-2, 2017, p. 41).

A key aspect of the Army Learning Concept is its ability to adapt and modify itself to meet student/mission needs (Department of the Army Pamphlet 600–25, 2015). Natural disasters or other unforeseen calamities in changing operational environments will require advanced use of a new mix of capabilities, formations, equipment, interfaces, and learning mediums. Another key strength of the Army Learning Concept is that it is not stuck on one specific instructional technique, instead, it focuses on key implementation methodologies to ensure better, long-term learning.

Use of enhanced technology, administered correctly, using key adult learning, student-centered principles are required with ALC to continually improve educational outcomes and improve speed/efficiencies (Department of the Army Pamphlet 600–25, 2015; Training and Doctrine Pamphlet 525-8-2, 2017). Although there are many research-based, stated benefits in the use of the Army Learning Concept, there are still some negative aspects of its use.

Overcoming Obstacles and Opposition to the Army Learning Concept/Model

Even with the stated importance and vital need of the Army Learning Concept, there are still those that oppose it due to issues relating to time and cost to implement (Ferguson, 2014). Yes, it will take time to update any Army courseware that isn't already using the Army Learning Concept and yes it does take time to present instructional content in a more dynamic and engaging way than a regular lecture. Yet, the results are worth the time invested in that soldiers are better educated, have higher motivation, and retain the information/knowledge longer (Himmel, 1970; Keller, 1987; Training and Doctrine Pamphlet 525-8-2, 2011). Although there are costs with redesigning any learning system as well as using new technologies, again the rewards are worth the investment. By improving training, student-soldiers will be better able to address current and future issues and will require less retraining due to improved long-term learning using the Army Learning Concept (Training and Doctrine Pamphlet 525-8-2, 2011). Additionally, one can argue that federal monies are currently being wasted in educational programs (such as Army Structured Self Development) and courses that do not fully implement the Army Learning Concept (Anders, 2018). Although some arguments do exist against the full implementation of the Army Learning Concept (investments of time and money), they hold little credence to its long-term outcomes.

Full and continued implementation of the Army Learning Model can only occur if appropriate policies are established and enforced that allow for rapid education adaption and the establishment of a culture that values and celebrates the power of education, training, and lessons learned as a force multiplier (Anders, 2018). Students at all levels need to be presented with instruction that is challenging, impactful, experiential, realistic, current, and future-leaning. This instruction must integrate critical thinking and situational awareness so that soldiers will be well equipped to deal with dynamic threats that don't always follow Army doctrine. This type of training is also vital to ensure that student-soldiers truly learn the information for the long-term and fully turn data into usable knowledge (Doctrine Pamphlet 525-8-2, 2017).

Time is an ongoing issue within the Army, but we cannot afford to waste time by forcing soldiers to sit through ineffective death by PowerPoint/slide-show presentations so that administrative leadership can check the box to say that training has been completed (Anders, 2018; Training and Doctrine Pamphlet 525-8-2, 2011). Proper use of instructional technology and implementation of the Army Learning Concept means that we would eliminate needless classes and ineffective instructional techniques such as the demotivating, unengaging pure lecture format and instead have life-long, student-centered, motivational, critical thinking, hands-on, educational events (Training and Doctrine Pamphlet 525-8-2, 2017).

CHAPTER SEVEN: CONCLUSION

This book has worked to express the powerful ideals of The Army Learning Concept (ALC) and the Army Learning Model (ALM), (*motivation, interaction, critical thinking,* and *experiential learning*), and why its use and implementation must occur to benefit all students, instructors/facilitators, and the educational experience itself. It must be remembered that ALM is not just a model/technique but is an ongoing learning environment that is adaptive and provides a continuum of life-long educational opportunities to deliver knowledge and skills to students under multiple conditions. ALC accomplishes this by using proven effective educational methodologies and new technologies (Training and Doctrine Pamphlet 525-8-2, 2017). Yet for the Army Learning Concept to be successful, students must fulfill their vital requirements to prepare by being lifelong learners and able to both internalize and synthesize gained knowledge (Training and Doctrine Pamphlet 525-8-2, 2017). The needed way forward to ensure ALC success is for the Army and the Joint Force to stay on the cutting edge of educational technology and learning science to ensure operational agility and superiority to fulfill the Army Operating Concept (Department of the Army Pamphlet 600–25, 2015; NCO 2020 Strategy,

2016). "To develop human capital effectively and achieve the Army goals, the Army must emphasize the train-coach-mentor approach, embody "leaders teach, teachers lead," and manage specialists" (Training and Doctrine Pamphlet 525-8-2, 2017, p. 29). Similarly, higher academia and business must develop this type of mentality in order to continually prepare the leaders, innovators, and workforce of the future.

A FINAL NOTE FOR THOSE IN THE MILITARY

As military educational professionals, it is important to remember that the Army requires *leaders to teach and teachers to lead*:

> Army leaders will embark on a program to select, develop, and assess experienced professionals continuously from officer, warrant officer, NCO, and civilian grades assigned as teachers. This reinvestment will help the Army identify suitable teachers, positioning them in assignments that permit maturation, growth, and reward through recognition and promotion. These programs must include a continuing education requirement to ensure those selected maintain currency in their area of expertise. The program's intent is to produce superior teachers and training and education developers, grounded in the Army Profession,

> knowledgeable in emerging learning technologies, possessing sufficient expertise with balanced experience in institutional and operational assignments. (Training and Doctrine Pamphlet 525-8-2, 2017, p. 30)

This can only be achieved by a mindset of lifelong learning. We should all continue to pursue our civilian education, taking full advantage of all the programs the Army provides. We should seek out broadening assignments to enhance our experiences and learn new languages before and during deployment to make ourselves that much more operationally effective. We should always strive to set the example for civilians, NCOs, and Officers to follow, holding education and the continual pursuit of knowledge in the highest regard by always publicly recognizing educational accomplishments and motivating others to continue to improve as well.

ADDITIONAL RESOURCES

The following are a listing of great free resources that are available on the Internet dealing with education and technology to help keep you informed and empowered to maximize the educational experience through the use of the Army Learning Model.

COPYRIGHT

HARVARD LAW SCHOOL LIBRARY **Understanding & Finding Public Domain/Creative Commons Media:** explains copyright and provides many links to historical & newly created content. https://guides.library.harvard.edu/c.php?g=310751&p=2072820

IMAGES & VIDEOS

Pixabay: offers over 1 million free high-quality stock images & videos that are copyright free, under the Pixabay License, "safe to use without asking for permission or giving credit to the artist - even for commercial purposes." https://pixabay.com

Public Domain Vectors: offers copyright-free vector images in popular .eps, .svg, .ai and .cdr formats. You are free to edit, distribute and use the images for unlimited commercial purposes without asking permission. https://publicdomainvectors.org/en

Wikimedia Commons: part of creative commons, free-content Wikimedia family. Image & video available under licenses specified on description pages. https://commons.wikimedia.org/wiki/Commons:Free_media_resources

Generated Photos: 100,000 Faces generated by Artificial Intelligence (AI) free to download and use. All that is required is a link back to the site for attribution credit. https://generated.photos

Ted Talks: Short, enlightening, powerful videos on topics dealing with Technology, Entertainment and Design as well as much more (including education). https://www.ted.com/#

PODCASTS

EdSurge: weekly podcast about the future of learning and the intersection of technology & education. Host Jeffrey R. Young: frank/in-depth conversations with educators, tech innovators & scholars. https://player.fm/series/66740

Instructional Tech Talk: a clearinghouse offering multiple podcasts on a wide range of info about instructional tech: https://instructionaltechtalk.com

Teaching in Higher Ed: explores the art and science of being more effective at facilitating learning. Increase personal productivity, to have more peace and be more present for students. https://teachinginhighered.com/episodes

NEWS

The Tech Advocate: Provides a well-rounded, multi-faceted look at the past, present, the future of EdTech in the US and internationally, with the latest news and opinion. https://www.thetechedvocate.org/

Inside Higher Ed: leading digital media company in higher education. A great source for higher education news, thought leadership, & technology https://www.insidehighered.com/news/focus/technology

Ed Tech: explores technology and educational issues that leaders and educators face when evaluating and implementing new tech solutions https://edtechmagazine.com/higher/

VIDEO GAMES & SIMULATIONS

Digital Trends: How to make a video game: provides links and descriptions on different ways to easily create your own video games (experience not required) https://www.digitaltrends.com/gaming/how-to-make-a-video-game

Center for Online Education: 50 sites for free serious educational games/simulations https://www.onlinecolleges.net/50-great-sites-for-serious-educational-games

Steam: digital distribution service platform for video games and simulations. Free and paid games/simulations available for multiple platforms https://store.steampowered.com

PLAGIARISM/CHEATING

Plagiarism.org: a very informative site dealing with the issue of preventing and detecting plagiarism https://www.plagiarism.org

PlagTracker: free plagiarism checking for uploaded documents, will work with several different languages. https://www.plagtracker.com

turnitin **Turnitin:** the most popular plagiarism checker and analyzer. Also built into the Canvas Learning Management System (LMS). https://www.turnitin.com

VIRTUAL & AUGMENTED REALITY

VR360 **VR360:** news about educational uses and new technologies dealing with virtual reality and augmented reality. https://www.virtualreality-news.net

AR POST **ARPost:** virtual reality and augmented reality information about news, events, and apps. https://arpost.co

CLASSVR® **ClassVR:** provides the latest information on successful use of VR & AR in schools. https://www.classvr.com/school-virtual-reality-news

REFERENCES

Abele, A. (1995) Stimmung und Leistung [Mood and achievement]. Göttingen: Hogrefe.

Ahmady, G., & Ruhi, N. N. (2016). Impact of promoting whole-class discussions in geometry instruction on students' reasoning ability. *International Journal of Humanities and Cultural Studies* (IJHCS) ISSN 2356-5926, 1(1).

AITS. (2014). *Army intelligence training strategy.* Headquarters Department of the Army, Washington, DC.

Aji, C. A., & Khan, M. J. (2019). The impact of active learning on students' academic performance. *Open Journal of Social Sciences, 7*(3), 204-211.

Allen, C., & Gerras, S. (2009). Developing creative and critical thinkers. *Military Review, November/December*. Retrieved from https://apps.dtic.mil/dtic/tr/fulltext/u2/a515842.pdf

Alex, K., Boris, G., Vladimir, S., and Suzanne, M. (2003). Vygotsky's educational theory in cultural context. Cambridge, NY: Cambridge University Press.

Alexander, N., & LeBaron, M. (2009). Death of the role-play. C. Honeyman, J. Coben, and G. De Palo (Eds.), *Rethinking negotiation teaching: innovations for context and culture*. St Paul, MN: DRI Press, 2009. pp. 179-197.

Anders, B. (2011). Negative learning and its mitigation in Army simulations. In S. Hai-Jew (Ed.), *Virtual immersive and 3D Learning Spaces: Emerging technologies and trends.* (pp. 333-335). Hershey, PA: IGI Global.

Anders, B. (2014). More than just data: The importance of motivation, examples, and feedback in comprehending and retaining digital information. In S. Hai-Jew (Ed.), *Packaging digital information for enhanced learning and analysis: Data visualization, spatialization, and multidimensionality.* (pp. 37-46). Hershey, PA: IGI Global.

Anders, B. (2015). Use of video to enhance education. In S. Hai-Jew (Ed.), *Design strategies and innovations in multimedia presentations*. (pp. 189-201). Hershey, PA: IGI Global doi:10.4018/978-1-4666-8696-0.

Anders, B. (2017). *How to enhance instructional presence: Research & experience based techniques to improve both online & face-to-face instruction*. Manhattan, KS: Sovorel Publishing.

Anders, B. (2018). Exploratory study of the structured self development experience of enlisted soldiers in the Kansas Army National Guard (Doctoral dissertation). Kansas State University, Manhattan, KS.

Andersson, V., Balslev, H., & Schaltz, T. (2019). Sharing experiences and the co-creation of knowledge through personal stories–tools for critical thinking; student perspectives. *Innovative Practice in Higher Education.*

Ameerbakhsh, O., Maharaj, S., Hussain, A., & McAdam, B. (2019). A comparison of two methods of using a serious game for teaching marine ecology in a university setting. *International Journal of Human-Computer Studies, 127*, 181-189.

AO Foundation. (2014). Running a practical exercise. Arbeitsgemeinschaft für Osteosynthesefragen. Retrieved from https://www.google.com/url?sa=t&rct=j&q=&esrc=s&source=web&cd=1&ved=2ahUKEwjToY-ntp7lAhVBUhoKHfciCrIQFjAAegQIAxAB&url=https%3A%2F%2Fimages.aofoundation.org%2FCIP%2Fasset%2Fdownload%2FGallery%2F6566&usg=AOvVaw1ziubkAMsfhjBClsoX7ika

Apedoe, X. S., Ellefson, M. R., & Schunn, C. D. (2012). Learning together while designing: Does group size make a difference? *Journal of Science Education and Technology, 21*(1), 83-94.

Apperson, J. M., Laws, E. L., & Scepansky, J. A. (2008). An assessment of student preferences for PowerPoint presentation structure in undergraduate courses. *Computers & Education, 50*(1), 148-153.

Argyris, C. (1991). Teaching smart people how to learn. *Harvard business review, 69*(3).

Army Doctrine Publication 5-0. (2012). *The Operations Process.* Headquarters Department of the Army, Washington, DC.

Army Doctrine Publication 7-0. (2012). *Training units and developing leaders.* Headquarters Department of the Army, Washington, DC.

Army Doctrine Reference Publication 2-0. (2012). *Intelligence.* Headquarters Department of the Army, Washington, DC.

Army Master Resiliency Training. (2014). Master resilience training participant guide. *U.S. Army Comprehensive Soldier and Family Fitness: MRT Version* 3.1.

Army Regulation 25-50. (2013). *Preparing and managing correspondence.* Headquarters Department of the Army, Washington, DC.

Army Regulation 350-2. (2015). *Operational environment and opposing force program.* Headquarters Department of the Army, Washington, DC.

Army Basic Instructor Courseware. (2012). Army basic instructor course. *Department of the Army, Headquarters, United States Army Training and Doctrine Command.* Fort Monroe, VA: US Army.

Arpaci, I., Baloglu, M., Kesici, S. (2019). A multi-group analysis of the effects of individual differences in mindfulness on nomophobia. *Information Development 35*(2), 333-341.

Baily, L. (2015, January). When simulation goes too far… dangers become real. *Healthy Simulation.* Retrieved from https://www.healthysimulation.com/6950/when-simulation-goes-too-far-dangers-becomes-real

Bambaeeroo, F., & Shokrpour, N. (2017). The impact of the teachers' non-verbal communication on success in teaching. *Journal of advances in medical education & professionalism, 5*(2), 51.

Barnieu, J., Morath, R., Bryson, J., Hyland, J., Tucker, J. S., & Burnett, S. (2016). *Using Technology to Support the Army Learning Model*. ICF International Fairfax United States.

Barrows, H. S., & Tamblyn, R. M. (1980). *Problem-based learning: An approach to medical education*. New York, NY: Springer Publishing Company Inc.

Behling, D. U., & Williams, E. A. (1991). Influence of dress on perception of intelligence and expectations of scholastic achievement. *Clothing and Textiles Research Journal, 9*(4), 1-7.

Berg, R., Brand, A., Grant, J., Kirk, J., & Zimmerman, T. (2014). Leveraging recorded mini-lectures to increase student learning. *Online Classroom, 14*(2), 5-8.

Bernard, R. M., Abrami, P. C., Borokhovski, E., Wade, C. A., Tamim, R. M., Surkes, M. A., & Bethel, E. C. (2009). A meta-analysis of three types of interaction treatments in distance education. *Review of Educational Research, 79*(3), 1243-1289.

Bodner, G. M. (2001). Why lecture demonstrations are 'exocharmic' for both students and their instructors. *University Chemistry Education, 5*(1), 31-35.

Boucher, J. (2019). Top 10 gen Z statistics from 2018. CGK Center for Generational Kinetics. Retrieved from https://genhq.com/top-10-ways-gen-z-is-shaping-the-future

Brock, S., & Joglekar, Y. (2011). Empowering PowerPoint: Slides and teaching effectiveness. *Interdisciplinary Journal of Information, Knowledge, and Management, 6*(1), 85-94.

Brookfield, S. (1995). Adult learning: An overview. *International encyclopedia of education, 10*, 375-380.

Bourdeaux, R. & Schoenack, L. (2016). Adult student expectations and experiences in an online learning environment. *The Journal of Continuing Higher Education, 64*: 152-161, 2016.

Campbell, L. O., Heller, S., & DeMara, R. F. (2019). Implementing student-created video in engineering: An active learning approach for exam preparedness. *International Journal of Engineering Pedagogy, 9*(4).

Carlisle, V. J. (2016). *Understanding the effects of personal responsibility and environment on the development of self-directed learning: an exploratory study* (Doctoral dissertation, Kansas State University).

Caspi, A., Gorsky, P., & Chajut, E. (2003). The influence of group size on nonmandatory asynchronous instructional discussion groups. *The Internet and Higher Education, 6*(3), 227-240.

Cervantes, E. P. (2009). Livening up college English classes with games. In *English Teaching Forum Vol. 47*, No. 3, (p. 20). U.S. Department of State, Bureau of Educational and Cultural Affairs, Office of English Language Programs.

Chaney, W. R. (2005). Top-of-hour break renews attention span. *The Teaching Professor*, 19(6), 1-5.

Chew, A., & Kwok, D. (2014). Students' perceptions of online learning environment and its influence on test achievement. Paper presented at TLHE 2014: International Conference on Teaching and Learning in Higher Education, National University of Singapore. Retrieved from http://www.cdtl.nus.edu.sg/Tlhe/tlhe2014/abstracts/achew.pdf

Clapper, T. C. (2010). Role play and simulation. *The Education Digest, 75*(8), 39.

Cole, T., Cairns, P., & Gillies, M. (2015, October). Emotional and functional challenge in core and avant-garde games. *In Proceedings of the 2015 Annual Symposium on Computer-Human Interaction in Play* (pp. 121-126). ACM.

Cortright, R. N., Collins, H. L., Rodenbaugh, D. W., & DiCarlo, S. E. (2003). Student retention of course content is improved by collaborative-group testing. *Advanced Physiology Education 27*: 102–108.

Covington, M. V. (2000). Goal theory, motivation, and school achievement: An integrative review. *Annual review of psychology, 51*(1), 171-200.

Couwels, J. (2010, December). High-tech war games help save lives. *CNN*. Retrieved from http://www.cnn.com/2010/TECH/innovation/12/11/military.simulation/index.html

Cross, K. P. (1999). *Learning is about making connections: The cross papers number 3*. Mission Viejo, CA: League for Innovation in the Community College and Educational Testing Service.

Csikar, E., & Stefaniak, J. E. (2018). The utility of storytelling strategies in the biology classroom. *Contemporary Educational Technology, 9*(1).

Cuevas, J. (2015). Is learning styles-based instruction effective? A comprehensive analysis of recent research on learning styles. *Theory and Research in Education, 13*(3), 308-333.

Daluba, N. E. (2013). Effect of demonstration method of teaching on students' achievement in agricultural science. *World Journal of Education, 3*(6), 1-7.

Dave, R.H. (1975). Psychomotor levels. In R. J. Armstrong (Ed.), *Developing and writing behavioral objectives*. Tucson, AZ: Educational Innovators Press.

Davis, B. G. (2009). *Tools for teaching*. San Francisco: CA. Jossey-Bass Inc.

Department of the Army Pamphlet 600–25. (2015). *U.S. Army noncommissioned officer professional development guide*. Headquarters Department of the Army, Washington, DC.

Department of the Army Pamphlet 600–67. (1986). *Effective writing for Army leaders*. Headquarters Department of the Army, Washington, DC.

Dewey, J. (1910). *How we think*. New York, NY: D.C. Heath & Co.

DiCarlo, S. (2009). Too much content, not enough thinking, and too little FUN! *Advanced Physiology Education 33*: 257–264.

Di Stefano, G., Gino, F., Pisano, G. P., & Staats, B. R. (2016). Making experience count: The role of reflection in individual learning. *Harvard Business School NOM Unit Working Paper,* (14-093), 14-093.

Doran, G. T. (1981). There's a SMART way to write management's goals and objectives. *Management review, 70*(11), 35-36.

Douglas, K. M., Wells, M. H., Deschepper, E. J., & Donaldson, M. E. (2017). Traditional lecture versus video/discussion-based instruction and their effects on learning behavior guidance techniques. *Journal of Education and Ethics in Dentistry, 7*(2), 30.

Duran, D. (2017). Learning-by-teaching. Evidence and implications as a pedagogical mechanism. *Innovations in education and teaching international, 54*(5), 476-484.

Dweck, C.S. (1986). Motivational processes affecting learning. *American Psychologist, 41,* 1040- 1048.

Elder, L., Paul, R. (2016). The thinkers guide to analytic thinking. Tomales, CA: Foundation for Critical Thinking Press.

Facione, P. (2011). Critical thinking: What it is and why it counts. *Insight assessment, 2007*(1), 1-23.

Farashahi, M., & Tajeddin, M. (2018). Effectiveness of teaching methods in business education: A comparison study on the learning outcomes of lectures, case studies and simulations. *The International Journal of Management Education, 16*(1), 131-142.

Federation of American Scientists. (2006). Harnessing the power of video games for learning. Summit on Educational Games, Washington, D.C.

Feinberg, M., & Nemeth, C. (2008). The" rules" of brainstorming: An impediment to creativity? *California Digital Library, University of California.*

Ferguson, Keith H. (2014, November). The Army learning model: An example of cognitive dissonance. *ARMY Magazine.*

Field Manual 6-0. (2015). *Commander and staff organization and operations.* Headquarters Department of the Army, Washington, DC.

Flanagan, S. (2015). How does storytelling within higher education contribute to the learning experience of early years students? *The Journal of Practice Teaching and Learning, 13*(2-3), 162-184.

Fraser, K., Huffman, J., Ma, I., Sobczak, M., McIlwrick, J., Wright, B., & McLaughlin, K. (2014). The emotional and cognitive impact of unexpected simulated patient death: a randomized controlled trial. *Chest, 145*(5), 958-963.

Frenzel, A.C., Goetz, T., Lüdtke, O., Pekrun, R. & Sutton, R.E. (2009). Emotional transmission in the classroom: Exploring the relationships between teacher and student enjoyment. *Journal of Educational Psychology, 101*(3), 705-716. http://dx.doi.org/10.1037/a0014695

Gage, F. H. (2002). Neurogenesis in the adult brain. *Journal of Neuroscience, 22*(3), 612-613.

Gajjar B, Mirza N, Shah N, Mistry M. (2017). Assignment-based learning of essential medicine list concepts among undergraduate medical students. *National Journal of Physiology, Pharmacy and Pharmacology*, 2017;8:924-927.

Garrett, N. (2016). How do academic disciplines use PowerPoint? *Innovative Higher Education, 41*(5), 365-380.

Garrison, D. R., Anders, T., & Archer, W. (1999). Critical inquiry in a text-based environment: Computer conferencing in higher education. *The Internet and Higher Education,* 2(2), 87-103.

Garrison, D. R., Anderson, T., & Archer, W. (2001). Critical thinking, cognitive presence, and computer conferencing in distance education. *American Journal of distance education, 15*(1), 7-23.

Gedera, D., Williams, J., & Wright, N. (2015). Identifying factors influencing students' motivation and engagement in online courses. *Motivation, Leadership and Curriculum Design* (pp. 13-23).

Gibson, D. (Ed.). (2006). *Games and Simulations in Online Learning: Research and Development Frameworks*. Hershey, PA: IGI Global.

GlobeNewswire. (2014, January). Majority of Americans prefer hands-on training in educational settings, survey finds. *West*. Retrieved from https://www.globenewswire.com/news-release/2014/01/29/605590/10065857/en/Majority-of-Americans-Prefer-Hands-On-Training-in-Educational-Settings-Survey-Finds.html

Gonzalez, J. (2019, July). Think twice before doing another historical simulation. *Cult of Pedagogy*. Retrieved from https://www.cultofpedagogy.com/classroom-simulations/?utm_source=feedburner&utm_medium=twitter&utm_campaign=Feed%3A+CultOfPedagogy+%28Cult+of+Pedagogy%29

Grange, C., & Miller, A. (2018). Teaching introduction to Psychology: Promoting student learning using digital storytelling and community engagement. *International Journal of Teaching and Learning in Higher Education, 30*(1), 172-183.

Greene, H., & Crespi, C. (2012). The value of student created videos in the college classroom-an exploratory study in marketing and accounting. *International Journal of Arts & Sciences, 5*(1), 273.

Gudonienė, D., & Rutkauskienė, D. (2019). Virtual and augmented reality in education. *Baltic journal of modern computing, 7*(2), 293-300.

Gunter, G., Kenny, G., & Junkin, S. (2018). The narrative imperative: Creating a storytelling culture in the classroom. In *Educational Technology and Narrative* (pp. 21-38). Cham, Switzerland: Springer.

Guo, P. J., Kim, J., & Rubin, R. (2014, March). How video production affects student engagement: An empirical study of MOOC videos. *In Proceedings of the first ACM conference on Learning@ scale conference* (pp. 41-50). ACM.

Gurney, D. J., Howlett, N., Pine, K., Tracey, M., & Moggridge, R. (2017). Dressing up posture: The interactive effects of posture and clothing on competency judgements. *British journal of Psychology, 108*(2), 436-451.

Harland, T. (2003). Vygotsky's zone of proximal development and problem-based learning: Linking a theoretical concept with practice through action research. *Teaching in higher education, 8*(2), 263-272.

Hammen, C. S., & Kelland, J. L. (1994). Attendance and grades in a human physiology course. *Advanced Physiology Education 12*: 105–108.

Harris, M. (2014, September). *College teaching with games. For the win!* Higher Ed Professor: Demystifying Higher Education. Retrieved from http://higheredprofessor.com/2014/09/04/college-teaching-with-games

Harrow, A.J. (1972). *A taxonomy of the psychomotor domain*. New York, NY: David McKay Co.

Hascher, T. (2010). Learning and Emotion: perspectives for theory and research. *European Educational Research Journal, 9*(1), 13-28.

Himmel, C. (1970). A critical review and analysis of self-directed learning methods utilized in the teaching of undergraduate psychology courses (Doctoral dissertation). Champaign, IL: University of Illinois at Urbana-Champaign.

Hoque, M. E. (2016). Three domains of learning: Cognitive, affective and psychomotor. *The Journal of EFL Education and Research, 2*(2), 45-52.

Horne, A., & Rosdahl, J. (2017). Teaching clinical ophthalmology: Medical student feedback on team case-based versus lecture format. *Journal of Surgical Education, 74*(2), 329-332.

Inuwa, U., Abdullah, Z., & Hassan, H. (2018). A mixed-method study of the effect of the demonstration method on students' achievement in financial accounting. *International Journal of Instruction, 11*(4), 577-592.

Jackson, V. A., & Back, A. L. (2011). Teaching communication skills using role-play: An experience-based guide for educators. *Journal of palliative medicine, 14*(6), 775-780.

Jacobs, G. M., Renandya, W. A., & Power, M. (2016). *Simple, powerful strategies for student-centered learning*. New York, NY: Springer.

Jensen, E. (2000). Moving with the brain in mind. *Educational leadership, 58*(3), 34-38.

Johnson, J. A., & Ferguson, M. D. (2018). Setting the foundation for effective learning: Utilizing the cognitive, affective, and psychomotor domains to establish rigorous performance learning objectives in postsecondary aviation programs. *The Collegiate Aviation Review International, 16*(1).

Kadoun, D. (2015). Academic gaming in the classroom. *Journal on Best Teaching Practices, 2*(2), 18-20.

Kaltura, (2015). *The state of video in education: A Kaltura report*. Retrieved from https://site.kaltura.com/rs/984-SDM-859/images/The_State_of_Video_in_Education_2015_a_Kaltura_Report.pdf

Keller, J. M. (1987a). Development and use of the ARCS model of instructional design. *Journal of Instructional Development, 10*(3), 2-10. Retrieved from http://www.jstor.org/stable/pdfplus/30221294.pdf

Keller, J. M. (1987b). Strategies for stimulating the motivation to learn. *Performance+ Instruction, 26*(8), 1-7.

Keller, J. M. (1987c). The systematic process of motivational design. *Performance+ Instruction, 26*(9-10), 1-8.

Keller, J. M. (2008). What is the ARCS model? Retrieved from http://arcsmode.ipower.com/Mot%20dsgn%20A%20model.htm

Kempermann, G., Wiskott, L., & Gage, F. H. (2004). Functional significance of adult neurogenesis. *Current opinion in neurobiology, 14*(2), 186-191.

Kizil, M. S., Hancock, M. G., & Edmunds, O. T. (2001). Virtual reality as a training tool. *In Proceedings of the Australian Institute of Mining and Metallurgy Youth Congress* (pp. 9-12).

Knowles, M. (1973). *The Adult learner: A neglected species*. Houston, TX: Gulf Publishing Company.

Knowles, M. (1984). Andragogy in action: Applying modern principles of adult learning. San Francisco, CA: Jossey-Bass.

Knowles, M. Holton, E., & Swanson, R. (2005). *The adult learner: The definitive classic in adult education and human resource development*. Burlington, MA: Elsvier.

Kodotchigova, M. A. (2002). Role play in teaching culture: Six quick steps for classroom implementation. *The Internet TESL Journal, 8*(7).

Kolb, A. & Kolb, D. (2005). Learning styles and learning spaces; Enhancing experiential learning in higher education. *Academy of Management Learning & Education. 2005, Vol. 4, No. 2*, 193-212.

Kolb, A. & Kolb, D. (2017). Experiential learning theory as a guide for experiential educators in higher education. *ELTHE: A Journal for Engaged Educators, Vol. 1*, No. 1, pp. 7–44.

Kolb, D. A., Boyatzis, R. E., & Mainemelis, C. (1999). Experiential learning theory: Previous research and new directions. In R. J. Sternberg and L. F. Zhang (Eds.), *Perspectives on cognitive, learning, and thinking styles.* NJ: Lawrence Erlbaum, 2000.

Konok, V., Gigler, D., Bereczky, B., Miklosi, A. (2016). Humans' attachment to their mobile phones and its relationship with interpersonal attachment style. *Computers in Human Behavior, 61*, 537-547.

Krathwohl, D. R. (2002). A revision of Bloom's taxonomy: An overview. *Theory into practice, 41*(4), 212-218.

Krathwohl, D. R., & Anderson, L. W. (2009). *A taxonomy for learning, teaching, and assessing: A revision of Bloom's taxonomy of educational objectives*. New York, NY: Longman.

Krathwohl, D. R., Bloom, B. S., & Masia, B. B. (1964). *Taxonomy of educational objectives, the classification of educational goals; Handbook II: The affective domain.* New York, NY: David McKay.

Krokos, E., Plaisant, C., & Varshney, A. (2018). Virtual memory palaces: immersion aids recall. *Virtual Reality, 23*(1), 1-15.

Lateef, F. (2010). Simulation-based learning: Just like the real thing. *Journal of Emergencies, Trauma and Shock, 3*(4), 348.

Lean, J., Moizer, J., Towler, M., & Abbey, C. (2006). Simulations and games: Use and barriers in higher education. *Active learning in higher education, 7*(3), 227-242.

LeBlanc, V. R. (2019). The relationship between emotions and learning in simulation-based education. *Simulation in Healthcare: Volume 14* - Issue 3 - p 137–139.

Lele, A. (2013). Virtual reality and its military utility. *Journal of Ambient Intelligence and Humanized Computing, 4*(1), 17-26.

Lin, C. H., Liu, E. Z. F., Chen, Y. L., Liou, P. Y., Chang, M., Wu, C. H., & Yuan, S. M. (2013). Game-based remedial instruction in mastery learning for upper-primary school students. *Journal of Educational Technology & Society, 16*(2), 271-281.

Lovelace, K. J., Eggers, F., & Dyck, L. R. (2016). I do and I understand: Assessing the utility of web-based management simulations to develop critical thinking skills. *Academy of Management Learning & Education, 15*(1), 100-121.

Luckin, R., & Fraser, D. S. (2011). Limitless or pointless? An evaluation of augmented reality technology in the school and home. *International Journal of Technology Enhanced Learning, 3*(5), 510-524.

Maier, H. (2002). Role playing: Structures and educational objectives. *e-Journal of the International Child and Youth Care Network, 36*, JAN2002.

Major, C. H., & Palmer, B. (2001). Assessing the effectiveness of problem-based learning in higher education: Lessons from the literature. *Academic exchange quarterly, 5*(1), 4-9.

Majuri, J., Koivisto, J., & Hamari, J. (2018). Gamification of education and learning: A review of empirical literature. In *Proceedings of the 2nd International GamiFIN Conference, GamiFIN 2018*. Pori, Finland.

Marine Board. (1996). *Simulated voyages: using simulation technology to train and license mariners*. Washington, D.C.: National Academies Press.

Matlin, M. (2003). *Cognition*. Hoboken, NJ: Wiley & Sons.

McCabe, D. L., Butterfield, K. D., & Treviño, L. K. (2012). *Cheating in college: Why students do it and what can be done about it.* Baltimore, MD: John Hopkins University Press.

McDermott, L. (1993). How we teach and how students learn. *Academic Science 701*: 9 –20, 1993.

McIsaac, M. S., Blocher, M. J., Mahes, V., & Vrasidas, C. (2006). Student and teacher perceptions of interaction in online computer-mediated communication. *Educational Media International, 36*(2), 121–131.

McKeachie, W. (1994). *Learning and cognition in the college classroom. Teaching tips: strategies, research and theory for college and university teachers.* Lexington, MA: Health.

McNeill, J. H., & Payne, P. K. (1952). Cooperative learning groups at the college level. *Educational Leadership, 54.*

McSharry, G., & Jones, S. (2000). Role-play in science teaching and learning. *School Science Review, 82*, 73-82.

Merieb, E. N. & Hoehn, K. (2007). *Human anatomy & physiology, 7th edition*. United Kingdom: Pearson International Edition.

Merriam, S. (2001). Andragogy and self-directed learning: Pillars of Adult Learning Theory. *New Directions for Adult and Continuing Education, no. 89*. Spring 2001. Hanover, PA: Jossey-Bass.

Mihailidis, P. (2013). A tethered generation: Exploring the role of mobile phones in the daily life of young people. *Mobile Media & Communication 2*(1), 58-72

Miller, G. E. (1962). An inquiry into medical teaching. *Journal of Medical Education 37*: 185–191.

Miller, M. M., & Brannon, L. A. (2017). Testing mindfulness-based acceptance against implementation intentions to discourage counterintentional cell phone use. *Mindfulness, 8*(5), 1212-1224.

Montgomery, K., Brown, S., & Deery, C. (1997). Simulations: Using experiential learning to add relevancy and meaning to introductory courses. *Innovative Higher Education, 21*(3), 217-229.

Moore, M. G. (1989). Three types of interaction. *American Journal of Distance Education, 3*(2), 1-6.

Moore, M. G. and Thompson, M. M. (1997). The effects of distance learning: Revised edition. *ACSDE Research Monograph, 15,* Penn State University, PA.

Morena, M. A., Smith, S., & Talbert, R. (2019). Video made the calculus star. *PRIMUS, 29*(1), 43-55.

Morris, C. (2018). *Why Walmart and other F500 companies are using virtual reality to train the next generation of American workers.* Retrieved from https://www.cnbc.com/2018/10/29/why-f500-companies-use-virtual-reality-to-train-workers-of-the-future.html

Mosca, J. B., Curtis, K. P., & Savoth, P. G. (2019). New approaches to learning for generation Z. *Journal of Business Diversity, 19*(3).

Murphy, P., Wilkinson, I., & Soter, A. (2016). Instruction based on discussion. In R. Mayer, & P. Alexander (Eds.). *Handbook of research on learning and instruction* (382-407). Abingdon, UK: Taylor & Francis.

Naqvi, S. (2015). Student created digital video and language learning: Voices from Omani classrooms. *In Third 21st CAF Conference at Harvard, in Boston, MA* (Vol. 6, No. 1).

Nestel, D., & Tierney, T. (2007). Role-play for medical students learning about communication: Guidelines for maximising benefits. *BMC medical education, 7*(1), 3.

NCO 2020 Strategy. (2016). NCO 2020 strategy: NCOs operating in a complex world. *Department of the Army, Headquarters, United States Army Training and Doctrine Command.* Fort Monroe, VA: US Army.

Nilson, L. B. (2016). *Teaching at its best: A research-based resource for college instructors.* San Francisco, CA: John Wiley & Sons.

O'Bannon, B. W., & Thomas, K. M. (2015). Mobile phones in the classroom: Preservice teachers answer the call. *Computers & Education, 85*, 110-122.

O'Donovan, B., Rust, C., & Price, M. (2016). A scholarly approach to solving the feedback dilemma in practice. *Assessment & Evaluation in Higher Education, 41*(6), 938-949.

OFCOM. (2006). *Television access services: Review of the code and guidance.* Office of Communications. Retrieved from https://www.ofcom.org.uk/__data/assets/pdf_file/0016/42442/access.pdf

Papargyris, A., & Poulymenakou, A. (2005). Learning to fly in persistent digital worlds: The case of massively multiplayer online role playing games. *ACM SIGGROUP Bulletin, 25*(1), 41-49.

Pekrun, R., Goetz, T., Frenzel, A. C., Barchfeld, P., & Perry, R. P. (2011). Measuring emotions in students' learning and performance: The Achievement Emotions Questionnaire (AEQ). *Contemporary educational psychology, 36*(1), 36-48.

Penner J. (1984). *Why Many College Teachers Cannot Lecture.* Springfield, IL: Thomas.

Peterson, T. O. (2004). So you're thinking of trying problem based learning?: Three critical success factors for implementation. *Journal of Management Education, 28*(5), 630-647.

Piaget, J. (1985). *The equilibration of cognitive structures*. Chicago, IL: University of Chicago Press.

Pierson, D. (2017). Reengineering Army education for adult learners. *Journal of Military Learning, 6*(11).

Qualters, D. (2002). Do students want to be active learners. *Journal of the Scholarship of Teaching and Learning, 51*-60.

Quinkert, K. A., Morrison, J. E., Fletcher, J. D., Moses, F. L., & Roberts, E. J. (2007). The Army science of learning workshop. *ARI Research Note 2007-02*. Arlington, VA: US Army Research Institute for the Behavioral and Social Sciences (DTIC No. ADM001922).

RAFT. (2013). Bridging the engagement gap with hands-on teaching. *Resource Area For Teaching*. Retrieved from https://raft.net/2017/05/09/case-for-hands-on-learning

Raju, P. K., & Sankar, C. S. (1999, January). Case study method of instruction in engineering classrooms. In *Forum of the South East Advanced Technological Education Consortium*, Auburn, AL.

Rao, D., & Stupans, I. (2012). Exploring the potential of role-play in higher education: Development of a typology and teacher guidelines. *Innovations in Education and Teaching International, 49*(4), 427-436.

Rao, S. P., DiCarlo, S. E. (2001). Active learning of respiratory physiology improves performance on respiratory physiology examinations. *Advanced Physiology Education 25*: 55– 61.

Reed, G. (2018). The rhetoric of character and implications for leadership. *The Journal of Character & Leadership Development, 5*(1), 5029.

Reedy, G. B. (2015). Using cognitive load theory to inform simulation design and practice. *Clinical Simulation in Nursing, 11*(8), 355-360.

Reinerman-Jones, L., Lameier, E., Biddle, E., & Boyce, M. W. (2017). *Informing the long-term learner model: motivating the adult learner (Phase 1)* (No. ARL-TR-8160). US Army Research Laboratory Aberdeen Proving Ground United States.

Reivich, K. J., Seligman, M. E., & McBride, S. (2011). Master resilience training in the US Army. *American Psychologist, 66*(1), 25.

Reynolds, K. M., Roberts, L. M., & Hauck, J. (2017). Exploring motivation: Integrating the ARCS model with instruction. *Reference Services Review, 45*(2), 149-165. *University Libraries Faculty & Staff Contributions. 96*. Retrieved from https://scholar.colorado.edu/libr_facpapers/96

Reynolds, G. (2012). PresentationZen: Simple ideas on presentation design and delivery. Berkeley, CA: New Riders.

Rickes, P. C. (2016). Generation in flux: How Gen Z will continue to transform higher education space. *Planning for Higher Education, 44*(4), 21-45.

Rismark, M., & Sølvberg, A. M. (2019). Video as a learner scaffolding tool. *International Journal of Learning, Teaching and Educational Research, Vol. 18, No. 1,* pp. 62-75, January 2019, https://doi.org/10.26803/ijlter.18.1.5

Roberts, D. (2018). The engagement agenda, multimedia learning and the use of images in higher education lecturing: or, how to end death by PowerPoint. *Journal of Further and Higher Education, 42*(7), 969-985.

Rothman, T., Romeo, L., Brennan, M., & Mitchell, D. (2011). Criteria for assessing student satisfaction with online courses. *International Journal for e-Learning Security, 1*(1-2), 27-32.

Russell, C., & Shepherd, J. (2010). Online role-play environments for higher education. *British Journal of Educational Technology, 41*(6), 992-1002.

Salas, E., Rosen, M. A., Pavlas, D., Jensen, R., Fu, D., Ramachandran, S., ... & Lampton, D. R. (2009)

Samman, S., Moshell, M., Clark, B., Brahwaite, C. (2009). *Learning to decode nonverbal cues in cross-cultural interactions*. Arlington, VA: U.S. Army Research Institute for the Behavioral and Social Sciences.

Sargeant, S. (2012). 'I don't get it': A critical reflection on conceptual and practical challenges in teaching qualitative methods. *Psychology Learning & Teaching, 11*(1), 39-45.

Scherpereel, C. M., & Bowers, M. Y. (2014, February). Using critical problem based learning factors in an integrated undergraduate business curriculum: A business course success. In *Developments in Business Simulation and Experiential Learning: Proceedings of the Annual ABSEL conference* (Vol. 33).

Schürer-Necker, E. (1984) Das emotionale erregungspotential [The emotional arousal level], in A. von Eye & W. Marx (Eds) Semantische dimensionen. *Verhaltenstheoretische Konzepte einer psychologischen Semantik*. Göttingen: Hogrefe.

Schürer-Necker, E. (1994) *Gedächtnis und emotion. Zum einfluss von emotionen auf das behalten von texten* [Memory and emotion. The influence of emotions on text restoring skills]. Weinheim: Beltz, Psychologie Verlags Union.

Seaman, M. A. (1998). Developing visual displays for lecture-based courses. *Teaching of Psychology, 25*(2), 141–145.

Selwyn, N. (2016). Digital downsides: Exploring university students' negative engagements with digital technology. *Teaching in Higher Education, 21*(8), 1006-1021. https://doi.org/10.1080/13562517.2016.1213229

Serrato, M. (2016). 'Watch-Think-Write' and other proven strategies for using video in the classroom. *KQED: Professional Learning*. Retrieved from https://ww2.kqed.org/education/2016/08/23/watch-think-write-and-other-proven-strategies-for-using-video-in-the-classroom/

Shmaefsky, B. R. (2005). MOS: The critical elements of doing effective classroom demonstrations. *Journal of College Science Teaching, 35*(3), 44.

Shors, T. J., Anderson, M. L., Curlik II, D. M., & Nokia, M. S. (2012). Use it or lose it: how neurogenesis keeps the brain fit for learning. *Behavioural brain research, 227*(2), 450-458.

Shiroma, P. R., Massa, A. A., & Alarcon, R. D. (2011). Using game format to teach psychopharmacology to medical students. *Medical teacher, 33*(2), 156-160.

Simkin, M. G., & McLeod, A. (2010). Why do college students cheat?. *Journal of Business Ethics, 94*(3), 441-453.

Steinwachs, B. (1992). How to facilitate a debriefing. *Simulation & gaming, 23*(2), 186-195.

Straus, S. G., Lewis, M. W., Connor, K., Eden, R., Boyer, M. E., Marler, T., Carson, C. M., Grimm, G. E., & Smigowski, H. (2019). *Collective simulation-based training in the US Army*. Santa Monica, CA: RAND Corporation.

Stuart J. & Rutherford R. J. (1978). Medical student concentration during lectures. *Lancet 2*: 514 –516.

Subhash, S., & Cudney, E. A. (2018). Gamified learning in higher education: A systematic review of the literature. *Computers in Human Behavior, 87*, 192-206.

Sutcliffe, M. (2002). Simulations, games and role-play. *The handbook for economics lecturers*, 1-26.

Swanson, D. B., Case, S. M., Luecht, R. M., & Dillon, G. F. (1996). Retention of basic science information by fourth-year medical students. *Academic Medicine 71*: S80 –S82.

Taubert, M., Villringer, A., & Lehmann, N. (2015). Endurance exercise as an "endogenous" neuro-enhancement strategy to facilitate motor learning. *Frontiers in human neuroscience, 9*, 692.

Tawfik, A. A., Schmidt, M. M., & Msilu, F. (2018). Stories as decision scaffolds: Understanding nonlinear storytelling using case-based reasoning and educational design research. In *Educational Technology and Narrative* (pp. 21-38). Cham, Switzerland: Springer.

Taylor, P. J., Russ-Eft, D. F., & Chan, D. W. (2005). A meta-analytic review of behavior modeling training. *Journal of applied psychology, 90*(4), 692.

Temple, P. (2018, May). Why you should avoid white backgrounds. Master your message for maximum impact. Retrieved from http://presentationsforbusiness.com/professional-speaker-image/white-death-on-the-podium

Tisdell, C., & Loch, B. (2017). How useful are closed captions for learning mathematics via online video?. *International Journal of Mathematical Education in Science and Technology, 48*(2), 229-243.

Training Circular 25-20. (1993). *A leader's guide to after-actions reviews.* Washington, DC: Department of the Army.

Training and Doctrine Pamphlet 350-70-1. (2019). *Training development in support of the operational training domain.* Fort Eustis, VA: Department of the Army, Headquarters, United States Army Training and Doctrine Command.

Training and Doctrine Pamphlet 350-70-7. (2018). *Army learning: Army educational processes.* Fort Eustis, VA: Department of the Army, Headquarters, United States Army Training and Doctrine Command.

Training and Doctrine Pamphlet 350-70-14. (2015). *Training and education development in support of the institutional domain.* Fort Eustis, VA: Department of the Army, Headquarters, United States Army Training and Doctrine Command.

Training and Doctrine Pamphlet 525-8-2. (2011). *The U.S. Army learning concept for 2015.* Fort Monroe, VA: Department of the Army, Headquarters, United States Army Training and Doctrine Command.

Training and Doctrine Pamphlet 525-8-2. (2017). *The U.S. Army learning concept for training and education, 2020-2040.* Fort Eustis, VA: Department of United States Army Training and Doctrine Command.

Training and Doctrine Pamphlet 525-8-3. (2011). *U.S. Army Training Concept, 2012-2020*. Fort Monroe, VA: Department of United States Army Training and Doctrine Command.

Training and Doctrine Pamphlet 525-8-6. (2018). *Concept for cyberspace and electronic warfare operations, 2025-2040*. Fort Eustis, VA: Department of United States Army Training and Doctrine Command.

Trent, K. (2019). Motivating and educating millennials. *Journal of Military Learning,* April 2019.

Tur, B., Harstad, J., & Antonakis, J. (2018, July). Effect of Charisma in Informal Leadership Settings: The cases of TED and Twitter. In *Academy of Management Proceedings* (Vol. 2018, No. 1, p. 13242). Briarcliff Manor, NY 10510: Academy of Management.

United Language Group. (2019, October). The immersive experience of video game language learning. ULG Blog. Retrieved from https://unitedlanguagegroup.com/blog/immersive-experience-video-game-language-learning

Vygotsky, L., S. (1978). *Mind in society: The development of higher mental processes.* Cambridge, MA: Harvard University Press.

van Dongen, E. V., Kersten, I. H., Wagner, I. C., Morris, R. G., & Fernández, G. (2016). Physical exercise performed four hours after learning improves memory retention and increases hippocampal pattern similarity during retrieval. *Current Biology, 26*(13), 1722-1727.

Van Horn, M. M. (September, 2016). Big data war games necessary for winning future wars. Military Review Exclusive Online. Retrieved from https://www.armyupress.army.mil/Portals/7/Army-Press-Online-Journal/documents/Van-Horn.pdf

viar360. (2017, June). *VR training brings skills and confidence to professionals*. Retrieved from https://www.viar360.com/vr-training

Vlachopoulos, D., & Makri, A. (2017). The effect of games and simulations on higher education: a systematic literature review. *International Journal of Educational Technology in Higher Education, 14*(1), 22.

Vouloumanos, A., Kiehl, K., Werker, J., & Liddle, P. (2001). Detection of sounds in the auditory stream: Event-related fMRI evidence for differential activation to speech and nonspeech. *Journal of Cognitive Neuroscience 13:7.*

Warrick, P. (2013). The Art and Science of Teaching. *Hawker Brownlow Thinking & Learning Conference.* Melbourne, Australia, 18 May 2013.

Watson, G. R., & Sottile, J. (2010). Cheating in the digital age: Do students cheat more in online courses?. *Online Journal of Distance Learning Administration 13*.1

Weimer, M. (2009, November). Problem-based learning: Benefits and risks. Faculty Focus: Higher ED Teaching Strategies from MAGNA Publications. Retrieved from https://www.facultyfocus.com/articles/effective-teaching-strategies/problem-based-learning-benefits-and-risks

Wesch, M. (2017). *The art of being human*. Lexington, KY: CreateSpace Independent Publishing.

Whitley, B. E., Jr. (1998). Factors associated with cheating among college students: A review. *Research in Higher Education, 39*(3), 236-274.

Widder, B. (2017, May). How to make a video game (experience not required). *Digital Trends*. Retrieved from https://www.digitaltrends.com/gaming/how-to-make-a-video-game

Wieman, C. (2007). Why not try a scientific approach to science education? *Change: The Magazine of Higher Learning, 39*(5), 9-15.

Wiggins, B. E. (2016). An overview and study on the use of games, simulations, and gamification in higher education. *International Journal of Game-Based Learning (IJGBL), 6*(1), 18-29.

Wlodkowski, R. J., & Ginsberg, M. B. (2017). *Enhancing adult motivation to learn: A comprehensive guide for teaching all adults.* San Fransisco, CA: John Wiley & Sons.

Wolff, M., Wagner, M., Poznanski, S., Schiller, J., & Santen S. (2014). Not another boring lecture: Engaging learners with active learning techniques. *The Journal of Emergency Medicine, Vol. 48*, No. 1.

Wong, L., & Gerras, S. J. (2015). *Lying to ourselves: dishonesty in the Army profession*. Army War College Carlisle Barracks PA Strategic Studies Institute.

Wood, E., Mirza, A., & Shaw, L. (2018). Using technology to promote classroom instruction: Assessing incidences of on-task and off-task multitasking and learning. *Journal of Computing in Higher Education, 30*(3), 553-571.

Worthington, D., & Levasseur, D. (2015). To provide or not to provide course PowerPoint slides? The impact of instructor-provided slides upon student attendance and performance. *Computers & Education 85* (2015) 14-22.

Yong, J. (2019, August 7). New study shows where 'growth mindset' training works (and where it doesn't). *EdSurge: Education Research*. Retrieved from: https://www.edsurge.com/news/2019-08-07-new-study-shows-where-growth-mindset-training-works-and-where-it-doesn-t

Yu, H., Glanzer, P. L., Johnson, B. R., Sriram, R., & Moore, B. (2018). Why college students cheat: A conceptual model of five factors. *The Review of Higher Education, 41*(4), 549-576.

Zhang, X. C., Lee, H., Rodriguez, C., Rudner, J., Chan, T. M., & Papanagnou, D. (2018). Trapped as a group, escape as a team: applying gamification to incorporate team-building skills through an 'Escape Room'experience. Cureus, 10(3).

GLOSSARY

Active Listening: Consciously paying attention to someone that is talking by maintaining eye contact, giving appropriate feedback to indicate that they are being paid attention to (such as a head nod), and then providing a brief back as to what was understood (paraphrasing).

Active Voice: Consists of creating shorter sentences by emphasizing the actor of the sentence by placing the main subject before the verb; it is made up of the standard English sentence order of subject-verb-object (example: You will have up to 60 seconds in which to emplace the Combat Application Tourniquet™). This type of effective communication is beneficial in that it increases the speed of communication and assists the reader in better understanding the main point of the message (Army Regulation 25-50, 2013).

ADDIE Model: A developmental process model, often used by the U.S. Army, that stands for Analysis, Design, Development, Implementation, and Evaluation. This model supports the development of learning products to address instructional requirements, identifies specific learning objectives that the instruction intends to address, provides ongoing assessment and/or evaluation feedback, and gains efficiencies by providing observations that focuses resources on mandatory learning requirements.

Adult Learner: *(for the purposes of this document)* Anyone old enough to be in the military and has a prior knowledge base - experience in what is being learned. This individual could be more of an adult learner (continuum) based on a combination of rank, years of service, and level/quality of experience in the subject matter being learned.

Army Learning Model (ALM): adaptive, continuous learning model routinely improved to provide quality, relevant, and effective learning experiences through outcome-oriented instructional strategies that foster thinking, initiative, & provide operationally relevant context which extends learning beyond the learning institution in a career-long continuum of learning through the significantly expanded use of network technologies.

Artificial Intelligence (AI): a computer system(s) that is capable of performing advanced tasks that would generally need human intelligence. Example processes include visual perception, translation between languages, speech recognition, decision-making, and even content creation.

Asynchronous: (opposite of synchronous) Instructional technique that allows students the ability to access content when they want as opposed to having to access the content at a specific time/location. Used in online instruction allowing courses (and access to it) to function around an individual's schedule. An example would be allowing students to view an on-demand video, instead of having to watch the video live (as it happens).

Augmented Reality (mixed reality): provides an added layer of reality by superimposing data or digital artifacts into the view of an individual who is wearing an augmented reality device such as the Microsoft HoloLens or using a cellphone with Google lens. A Microsoft HoloLens example scenario would be you look at an engine and the augmented reality automatically labels each part for you. A Google lens example would be you hold up your cellphone camera to documents written in another language and it instantly superimposes an English translation.

Blooms Revised Taxonomy: A standardized framework to organize and categorize the different learning objectives. The six different levels within Bloom's Revised Taxonomy from simple to complex are 1. Remember (recognizing/identifying/recalling/retrieving), 2. Understand (interpreting, exemplifying, classifying, summarizing, inferring, comparing and explaining), 3. Apply (executing/implementing), 4. Analyze (differentiating, organizing, and attributing), 5. Evaluate (checking/critiquing), 6. Create (generating/planning/producing). Additionally all of these educational objectives can be further categorized within different knowledge dimensions: factual knowledge (terminology, details, symbols), conceptual knowledge (interrelationship information within a larger structure and how items function together), procedural knowledge (ways/methods/techniques), and metacognitive knowledge (process of thinking, awareness of one's self-own cognition - reflection).

Cognitive Load: The mental load placed on a learner, amount of brain/mental processing required by the student during the learner process.

Cognitive Overload: when an individual is overwhelmed with information (requiring excessive brain processing) due to too much data, the information being too complicated, too many things occurring at the same time or if too many tasks are required of the individual (learner) at the same time.

Cognitive Theory of Multimedia Learning: theory expressing that people learn content better when they are exposed to words & images (still or moving) than by words alone. Research also found that there is a greater educational impact when words are presented aurally in conjunction with associated images of the topic (not images of the words).

Constructivism: the notion that there is an unmistakable powerful connection between what a student is already knowledgeable about and/or has the ability to do, and the new information/skill that a student needs to learn. The student will have to "construct" new meaning/interpretation within their minds between what was already known and the new knowledge/learning experience.

Critical Thinking: The intellectually disciplined process of actively and skillfully conceptualizing, applying, analyzing, synthesizing, and/or evaluating information gathered from, or generated by, observation, experience, reflection, reasoning, or communication, as a guide to belief and action" (Training and Doctrine Pamphlet 525-8-2, 2017, p. 21). Moving beyond linear thought to more complex ways of thinking – seeking a greater understanding of how outside influences pertain to the situation at hand – using reflection and creative thinking while also considering second and third-order effects.

Deep Learning: this is an approach as well as an attitude to learning that requires the student to use critical thinking higher-level cognitive capabilities (metacognition) to fully comprehend issues, understand and solve problems, and appreciate processes to build upon previous knowledge for long term retention of what was newly learned.

Digital Storytelling: the presentation of a story via a variety of any digital means such as through audio, video, animation, web story with images, podcasts, games, virtual reality, etc. Implies the use of digital media to present a story with motivation and emotion.

Emojis: Popular emotion icons generally representing facial expressions. Emojis are popular visual options when trying to add emotion to textual writing via texting through instant messenger or other texting services via modern cellphones (examples: ☺, ☹).

Etiquette/Netiquette: Refers to proper and appropriate ways that individuals are to communicate and interact with one another, either face to face or online (netiquette).

Experiential Learning: a process of learning through active engagement with the learning subject (a real event/experience) followed by purposeful reflection on what occurred (fully understanding/grasping all that happened), then conceptualizing and thinking about the event (integrating with prior knowledge), proceed by active experimentation – practice (gaining skill mastery of what was learned), finally applying what was learned to a new experience (and then repeating the process, a continuous spiral of learning), (Kolb, & Kolb, 2017). It isn't enough to simply have an experience, the full process described must occur for learning to fully take hold.

Extrinsic Motivation: motivational elements outside of one's self (internal desires). An example would be I want to learn this just so that I can pass the test or just so I can get an A letter grade. Extrinsic Motivation is the opposite of intrinsic motivation.

Flipped Classroom: an educational technique with students read or view instructional material (articles/videos) at home (outside of class time) and then use the actual class time to discuss the material learned or go through assignments together. This is opposite of the traditional receive lecture at school and then do homework outside of class.

Gamification: the deliberate implementation of game-design into a course via the use of elements such as leader boards, badges, levels, points/scores, and/or presenting the course as more of an adventure/quest as opposed to a regular class.

Generation Z: Students categorized as being born from 2000 to 2019. They are often described as being deeply connected with technology since they born with computers and the Internet already part of everyday life.

Intrinsic Motivation: a skill that can be developed by a student that gives the ability for one to be self-motivated without regard to outside influences (or lack thereof). This allows for a student to be motivated even when the instructor/facilitator isn't motivating; the student maintains an internal drive and persistence to succeed in their educational endeavors. An example of Intrinsic Motivation would be learning something because it is interesting or just for the experience of learning as opposed to just to pass a test. This is the opposite of Extrinsic Motivation.

Learning Objective: Primary goal or mission of the learning endeavor (multiple learning objectives can be identified for any learning endeavor). This states what students will be able to do (action verb) by the end of the lesson/class after teaching activities have occurred (example: students will be able to *apply* a tourniquet to a casualty). Additional specifics and grading criterial (standards) are given in the grading *rubric* for the lesson/class. To assist in organizing and standardizing the language used in a learning objective, *Blooms Revised Taxonomy* should be utilized.

Learning Scaffolding: Instructional techniques/content/assistive tools to temporarily help learners until they can either become more independent learners or have caught up with the rest of the class. Often associated with just-in-time learning, where a student receives additional help/instruction to learn content that should have been previously learned. Learning Scaffolding can include techniques/tools such as: videos, text, presentations, quizzes, games, simulations, tutoring and/or artificial intelligence assistive tools.

Learning Science: The science and research pertaining to education, how students learn, interpret, retain, and experience information for knowledge acquisition.

Life-long Learner: One who continuously pursues knowledge throughout their life. Implies the ability to be a self-directed learner, continuously seeking both formal and informal, both civilian and military, educational experiences.

Metacognition: An awareness and control of one's own cognition (thoughts and mental processes); thinking about one's own thinking. Being able to properly reflect (*reflection*) is considered a vital element of metacognition.

Millennials: Students categorized as being born from 1989 to 1999. Often described as being more connected with technology since they grew up during the information age.

Motivation: *(for the purposes of this document)* The deliberate act of instilling a drive or reason for the student to want to learn (can also be the facilitation of awakening or directing students' own internal self-motivation). Keller's (1987) ARCS Model of motivation is recommended in maximizing motivation in the classroom. *ARCS: Attention, Relevancy, Confidence, and Satisfaction.*

Negative Learning: the accidental or unintentional learning of something or a process while in the course of learning something else.

Neurogenesis: a process where the brain (even an adult's) creates new neurons. Research has proven that because of this all students can continue to learn, even in late adulthood.

Operational Environment: a composite of the conditions, circumstances, and influences that affect the employment of capabilities and bear on the decisions of the commander (Joint Publication (JP) 3–0). Army Doctrine Publication (ADP) 3–0 and Army Doctrine Reference Publication (ADRP) 3–0 describe the Operational Environment in terms of eight operational variables: political, military, economic, social, information, infrastructure, physical terrain, and time (PMESII–PT); Army forces use these operational variables to understand and analyze the broad environment in which they are conducting operations.

Reflection: advanced mental action requiring thinking about what has occurred and/or one's own thoughts/ways of thinking (metacognition). Reflection is also part of the second step of the Army Experiential Learning Model (Publish & Process) as well as an important aspect of critical thinking. Reflection allows for deeper contemplation and learning.

Resiliency: deals with helping individuals become better able to bounce back from stressful situations and better overcome psychological adversity. Resiliency training consists of helping Soldiers become more self-aware of their thoughts/feelings, self-regulating with their impulses/emotions, being more optimistic (noticing positive things in life, self, and others), improving mental agility (flexible/perspective thinking), identifying character strengths (or areas to improve within themselves), and connection (establishing positive relationships).

Rubric: An organizational tool to assist in grading that presents the specific criteria for each portion of a student's grade; specific criteria & performance requirements for what a student must do in their assignment/assessment in order to achieve each grade level.

Self-Motivation: a skill that can be developed by a student that gives the ability for one to be intrinsically motivated without regard to outside influences (or lack thereof). This allows for a student to be motivated even when the instructor/facilitator isn't motivating; the student maintains an internal drive and persistence to succeed in their educational.

S.M.A.R.T. Goals: objectives that are **S**pecific, **M**easurable, **A**chievable/**A**ctionable, **R**elevant, and **T**imely (**T**ime-based/**T**imebound).

Syllabus: a document that presents information about the course. Provides an outline about the topics covered and additional information about the course procedures. An effective syllabus should have a course description, instructor(s)/facilitator(s) information, instructional resources, course calendar, assignment information, learning outcomes, grading procedure/grading scale, policies/procedures/standards, and legal statements.

Synchronous: (opposite of asynchronous) Instructional technique that requires students to access instructional content at the same time as other students. An example would be requiring students to view a video live (as it happens) as opposed to allowing students to view an on-demand video instead. An example of a synchronous course would be a face-to-face course (together live in person) or a course that is conducted via video conferencing where all students are required to join the video conferencing event.

Virtual Reality: a created digital reality that can be made into anything that one wants from abstract to realistic and can be used for entertainment as well as for education. Items within virtual reality can be made so that the viewer/player can interact with said items as they move through the virtual reality environment.

Zone of Proximal Development: "the distance between the actual development level as determined by independent problem solving and the level of potential development as determined through problem-solving under adult guidance or in collaboration with more capable peers" (Vygotsky, 1978, p. 86). Although designed dealing with children's cognitive development, this has also been applied to higher education, helping students succeed by challenging their capabilities without cognitive overload (Harland, 2003).

INDEX

W

Z

Feedback Request and Additional Information

I would like to personally thank you for reading this book. Please provide a review on Amazon, I would appreciate it. Contact me at Sovorel Publishing: www.sovorelpublishing.com/index.php/contact-us or contact@sovorelpublishing.com

Consider purchasing my other book, *How to Enhance Instructional Presence* available on Amazon. It deals with how to specifically make students feel more like a real learning member of an instructional community (of inquiry) and not "just a number." The book describes how strategic implementations of video can be used to enhance social, cognitive, and teaching presence to maximize instructional effectiveness, engagement, instructor approachableness, student achievement, and improve educational experiences.

Also consider visiting my educational blog Sovorel, dealing with higher education, focusing on learning, instructional methodologies, instructional technologies, and educational techniques: www.sovorelpublishing.com

ABOUT THE AUTHOR: Brent A. Anders Ph.D.

Anders has worked in higher education for over twenty years, concentrating on media and online instruction. Anders has extensively worked as an educational media consultant (video production, live webcasting, student experience/ engagement, and instructional technology), a professor, and as a course developer/instructional designer. Anders has also been serving in the U.S. Army for over 25 years, first as an airborne infantryman, then as a certificated military instructor. Anders is currently a Sergeant Major and works as a U.S. Army National Guard international instructor assisting the Kansas State Partnership Program with the government of Armenia. Anders also does periodic speaking events throughout the United States, and authors books, research articles, and blog posts dealing with education. These different experiences have allowed Anders to observe many different learning scenarios/factors dealing with higher education, allowing him to identify problematic issues and successful solutions.

For a full listing of Brent A. Anders' academic publications please visit https://ksu.academia.edu/BrentAnders

Made in the USA
Monee, IL
05 November 2026

67378969R00164